GIS for Water Resources and Watershed Management

GIS for Water Resources and Watershed Management

Edited by

John G. Lyon

Taylor & Francis
Taylor & Francis Group

LONDON AND NEW YORK

First published 2003 by Taylor & Francis
11 New Fetter Lane, London EC4P 4EE

Simultaneously published in the USA and Canada
By Taylor & Francis Inc,
29 West 35th Street, New York, NY 10001

Taylor & Francis is an imprint of the Taylor & Francis Group

© 2003 Taylor & Francis
except Chapters 1, 3, 5 & 6 © American Water Resources Association; Chapter 13 ©
Elsevier Science; Chapter 14 © American Society for Photogrammetry and Remote Sensing;
Chapters 2, 4, 7, 17 and 18, which originate from a US Government source and are in the
public domain.

Typeset in Times and Helvetica by Sans Serif Inc., Michigan, USA
Printed and bound in Great Britain by TJ International Ltd, Padstow, Cornwall

Every effort has been made to ensure the advice and information in this book is true and
accurate at the time of going to press. However, neither the publisher nor the authors
can accept any legal responsibility or liability for any errors or omissions that may be
made. In the case of drug administration, any medical procedure or the use of technical
equipment mentioned within this book, you are strongly advised to consult the manufacturer's
guidelines.

British Library Catloguing in Publication Data
A catalogue record for this book is available from the British Library.

Library of Congress Cataloging in Publication Data
A catalog record for this book has been requested.

ISBN 0-415-28607-7

Contents

About the Editor

Lyon was interested early on in water resources, watersheds, and wetlands and other moderately disturbed systems as places to evaluate the condition of native vegetation communities. This interest was honed in his youthful wanderings in the mountains and alpine systems of the Pacific Northwest, California and Nevada, and Alaska. Systematic study of the Lower Columbia River and wetlands in undergraduate work at Reed College in his native Oregon, and graduate work at the University of Michigan yielded Bachelor's and Master's theses and a Doctoral Dissertation devoted to Great Lakes wetlands and other natural systems. Eighteen years as a faculty member and ultimately a full Professor of Civil Engineering and Natural Resources at Ohio State University were further devoted to scholarly pursuits of these interests. A body of work included remote sensor and GIS evaluations of water resources and wetlands, development of field methods for wetlands identification, and other efforts on hydrology, soil, agriculture, riverine, and Laurentian Great Lakes systems.

Contributors

Bruce C. Atherton
Department of Food, Biological and Agricultural Engineering, Ohio State University, Columbus, OH 43210, USA

Eric M. Beamer
Skagit System Cooperative, LaConner, WA 98257, USA

Timothy J. Beechie
Skagit System Cooperative, LaConner, WA 98257, USA

Kenneth N. Brooks
Department of Forest Resources, University of Minnesota, 1530 North Cleveland Avenue, St. Paul, MN 55108, USA

Daniel L. Civco
Department of Natural Resources, University of Connecticut, Storrs, CT 06269, USA

Jeffrey D. Colby
Department of Geography, East Carolina University, Greenville, NC 27858-4353, USA

Brian L. Cosentino
Washington State Department of Fish and Wildlife, Olympia, WA 98501-1091, USA

James F. Cruise
Civil and Environmental Engineering Department, University of Alabama, Huntsville, AL 35899, USA

David A. Eash
U.S. Geological Survey, Water Resources Division, P.O. Box 1230, Iowa City, IA 52244, USA

Jurgen Garbrecht
U.S. Department of Agriculture, Agricultural Research Service, Grazinglands Research Laboratory, 7207 W. Cheyenne Street, El Reno, OK 73036, USA

Donald Garofalo
U.S. Environmental Protection Agency, Environmental Photographic Interpretation Center, National Exposure Research Laboratory/ORD, 12201 Sunrise Valley Drive, 555 National Center, Reston, VA 20192, USA

Andrés R. Garcia-Martinó
U.S. Department of Agriculture, U.S. Forest Service, Caribbean National Forest, Puerto Rico

David C. Goodrich
U.S. Department of Agriculture, Agricultural Research Service, Southwest Watershed Research Center, 2000 E. Allen Road, Tucson, AZ 85719, USA

Weihe Guan
Forest Information Technology, Weyerhaeuser Company, MS WWC 2B2, 33405 8th Ave. S., Federal Way, WA 98003, USA

D. Phillip Guertin
Watershed Resources Program, School of Renewable Natural Resources, University of Arizona, Tucson, AZ 85721, USA

Craig A. Harvey
PixSell, Inc., NASA John C. Stennis Space Center, Bldg. 2105, Stennis Space Center, MS 39529, USA

Robert E. Holman
North Carolina Water Resources Research Institute, North Carolina State University, Campus Box 7912, Raleigh, NC 27695-7912, USA

Tom Krill
John Deere Co., Fort Collins, CO 80524, USA

Sergio L. Lostal
South Florida Water Management District, 3301 Gun Club Road, West Palm Beach, FL 33406, USA

Ross S. Lunetta
U.S. Environmental Protection Agency, National Exposure Research Laboratory/ORD, MD-56, Research Triangle Park, NC 27711, USA

John G. Lyon
P.O. Box 71926, Las Vegas, NV 89170-1926, USA

David R. Maidment
Center for Research in Water Resources, University of Texas, Austin, TX 78712, USA

Lawrence W. Martz
Department of Geography, University of Saskatchewan, 9 Campus Drive, Saskatoon, Saskatchewan, S7N 5A5, Canada

Richard L. Miller
NASA Earth System Science Office, John C. Stennis Space Center, Stennis Space Center, MS 39529, USA

Scott N. Miller
U.S. Department of Agriculture, Agricultural Research Service, Southwest Watershed Research Center, 2000 E. Allen Road, Tucson, AZ 85719, USA

Pawel J. Mizgalewicz
Center for Research in Water Resources, University of Texas, Austin, TX 78712, USA

David R. Montgomery
Department of Geological Science and Quaternary Research Center, University of Washington, Seattle, WA 98195, USA

Christian Nolte
International Institute of Tropical Agriculture (IITA), Nigeria

Lloyd P. Queen
School of Forestry, University of Montana, Missoula, MT 59812, USA

Merrill K. Ridd
Department of Geography, University of Utah, Salt Lake City, UT 84112, USA

J. M. Salisbury
Spatial and Environmental Information Clearinghouse, 003 Life Sciences East, Oklahoma State University, Stillwater, OK 74078, USA

Frederick N. Scatena
U.S. Department of Agriculture, U.S. Forest Service, Caribbean National Forest, Puerto Rico

F.R. Schiebe
SST Development Group, Inc., 824 North Country Club Road, Stillwater, OK 74075, USA

Gabriel S. Senay
EROS Data Center, Sioux Falls, SD 57198, USA

Michiharu Shiiba
Disaster Prevention Research Institute, Kyoto University, Uji 611, Japan

Patrick J. Starks
U.S. Department of Agriculture, Agricultural Research Service, Grazinglands Research Laboratory, 7207 W. Cheyenne Street, El Reno, OK 73036, USA

Yasuto Tachikawa
Disaster Prevention Research Institute, Kyoto University, Uji 611, Japan

Takuma Takasao
Department of Civil Engineering, Kyoto University, Kyoto 606, Japan

Leslie J. Turner
People Soft Global Services, 100 Four Falls Corporate Center, Suite 515, West Conshohocken, PA 19428, USA

Prasad Thenkabail
Center for Earth Observation, Department of Geology and Geophysics, Yale University, 210 Whitney Avenue, New Haven, CT 06520, USA

D. A. Waits
SST Development Group, Inc., 824 North Country Club Road, Stillwater, OK 74075, USA

Andrew Ward
Department of Food, Biological and Agricultural Engineering, Ohio State University, Columbus, OH 43210, USA

Glenn S. Warner
Department of Natural Resources, University of Connecticut, Storrs, CT 06269, USA

W. Scott White
Department of Geography, University of Utah, Salt Lake City, UT 84112, USA

Donald C. Williams
U.S. Army Corps of Engineers, Mississippi Valley Division, Vicksburg, MS 39180, USA

Wayne L. Wold
Boise Cascade Corporation, P.O. Box 50, Boise, ID 83728-0001, USA

Acknowledgements

Chapter 1 is reproduced by kind permission of the American Water Resources Association from the *Symposium on Geographic Information Systems in Water Resources*.

Chapters 2, 4, 7, 17 and 18 originate from a US Government source and are in the public domain.

Chapter 3 is reproduced by kind permission of the American Water Resources Association from the Journal of the American Water Resources Association.

Chapter 5 is reproduced by kind permission of the American Water Resources Association from *GIS and Water Resources*.

Chapter 6 is reproduced by kind permission of the American Water Resources Association from the symposium proceedings *GIS and Water Resources*.

Figures 10.1 and 10.2 are in the public domain.

Figures 12.1, 12.3, 12.5 © by South Florida Water Management District

Chapter 13 is reprinted from *Aquatic Botany* **58**, Donald C. Williams and John G. Lyon, "Historical aerial photographs and a geographic information system (GIS) to determine effects of long-term water level fluctuations on wetlands along the St Mary's River, Michigan, USA", 363–378, ©1997 with kind permission of Elsevier Science – NL, Sara Burgerhartstraat 25, 1055 KV Amsterdam, The Netherlands.

Chapter 14 is reproduced with permission, the American Society for Photogrammetry and Remote Sensing. Lunetta R., Cosentino B., Montgomery D., Beamer E. and Beechie T. "GIS-Based Evaluation of Salmon Habitat in the Pacific Northwest". *Photogrammetric Engineering and Remote Sensing*, Vol 63 no. 10 (October 1997), 1219–1229.

Figures 15.2 and 15.3 are reproduced by kind permission of the American Water Resources Association from Colby, Jeffrey D. 1996, Physical Characterisation of the Navarro Watershed for Hydrologic Simulation, *GIS and Water Resources* (Proceedings of Symposium), pp 383–392.

Figure 16.3 is reprinted from *Remote Sensing of Environment* **53**, Miller R.L. and J.F. Cruise, "Effects of Suspended Sediments on Coral Growth: evidence from remote sensing and hydrologic modeling", 177–187, ©1995 with permission from Elsevier Science.

Figures 16.4, 16.5, 16.6 and 16.7 are reproduced by kind permission of American Water Resources Association from Water Resources Bulletin (now the Journal of the American Water Resources Association).

GIS for Water Resources and Watershed Management

John Grimson Lyon, Editor

The chapter authors and the Editor are pleased to bring you this volume of outstanding contributions to water resources research. These authors and others have pioneered and developed viable applications of Geographic Information Systems (GIS) and continue to do so. As a result, the current state of water resources and watershed modeling has been advanced by the use of GIS and allied technologies (Maidment and Djokic, 2000). This volume reports their accomplishments and shares their insight with others who may find these contributions valuable to their own work.

A number of GIS applications to water resources and watersheds have been completed over the years, and they nicely illustrate the potential of the technology. Many of these efforts have resulted from the need to address difficult-to-achieve project goals. GIS applications are applied due to the variability of the resources over time and space, and the number of variables that must be evaluated.

Geographic Information Systems are databases that usually have a spatial component to the storage and processing of the data. Hence, they have the potential to both store and create maplike products. They also offer the potential for performing multiple analyses or evaluations of scenarios such as model simulations.

Data are stored in multiple files. Each file contains data in a coordinate system that identifies a position for each data point or entry. Characteristics of the data point are stored as "attributes." A database of individual files is developed and the combined files may contain characteristics or attributes such as stream locations, topography, water or soil chemical sampling, management practices, ownership, biota, point sources, and any other data that can be collected and have meaning for the analysis.

The GIS supplies value by virtue of the detail of the database, and how these data can be used to address the application of interest. Each variable or "layer" in the GIS supports the application with data on the characteristics of water resources or watersheds.

An important capability of GIS is the simulation of physical, chemical, and biological processes using models. GIS can potentially be used with deterministic or complex models based on algorithms simulating processes, or they can be applied with statistical models. The requirement is that the model to be applied has the capability to take spatial and/or multiple file or "layer" data as input to computations.

Geographic Information Systems are also amenable to evaluations as to quality. They may be used to evaluate the quality characteristics of the product, and to determine the inherent accuracy (Lyon, 2001). Many methods can assist the assessment of accuracy, and these methods can be used in experiments.

Advanced applications utilize GIS, remote sensor, and Global Positioning Systems (GPS) technologies to better address the project goals. These advanced approaches make use of more complex models and GIS technologies, data collection along with GPS (Kennedy, 1996; Van Sickle 2001), and remote sensing data collection and analyses.

There are a number of strengths that GIS technologies bring to water resources research. GIS allows for improved database organization and storage. The objectives of many watershed studies include watershed segmentation, identification of drainage divides and the network of channels, characterization of terrain slope and aspect, catchment configuration, and routing of the flow of water. Obtaining these variables has been difficult to do from paper maps and aerial photographs. These traditional methods are subject to errors related to manual operations. The work using traditional measurement methods has also proved to be time-consuming.

Once the GIS system is developed, it is straightforward to produce needed data, make maps, and provide simple or complex displays of model results.

The GIS also allows for integration of other sources of data. Historical maps and aerial photos can be of assistance. The U.S. Geological Survey (USGS), the U.S. Department of Agriculture, and the National Archives and Records Service hold originals of maps and photos. Copies of these can be obtained, corrected for inconsistencies, and scanned into a computer file for analyses (Garofalo, 2001).

The USGS also creates and sells Digital Elevation Model (DEM) data sets for the U.S. These data are sets of point elevations on a regular sampling grid. These point elevations are organized by USGS 1:24,000 scale or smaller scale quadrangles. The use of DEM products and custom DEMs are the subject of much of the book, and others (Maidment and Djokic, 2000).

The above USGS products are available at low cost. They may also be acquired at no cost from the USGS address on the Internet.

These USGS products do have the potential for being inaccurate. This may be due to systematic or nonsystematic errors, old source quadrangle maps, or other problems. Normally, the user will make any corrections or modifications for the needs of a given project. Corrections and methods to make corrections in DEM data are an important issue in this book.

It is also possible to purchase enhanced USGS products and other products from vendors who develop the improved DEM products. These enhanced or "value-added" products are quite good, and they represent a considerable cost savings as compared to generating the data anew or correcting available USGS versions. Vendors may also have unique products, or products that can be further enhanced to meet user needs. Examples would include customizing an existing data set to provide only information for a given watershed that is found on several map sheets, or developing a fine detailed DEM by use of airborne LIDAR measurements.

Another, potentially valuable data set is U.S. Department of Agriculture Natural Resource Conservation Service (NRCS, formerly Soil Conservation Service) soils data. Several programs have resulted in general soils data stored in digital form, and products are available at different scales and levels of detail. For example, the STATSGO data sets are files of soil survey soil type and boundary data, and the chapters present some of the capabilities of these and similar databases.

A number of projects have used satellite remote sensor measurements. The resulting remote sensor data provide valuable inputs to GIS, due to the difficulty of access to research areas or the size of the study areas. Supplying data on watersheds over sizable areas has been a valuable use of satellite systems, and has been conducted in a variety of ecosystems including Florida, Washington, the Great Lakes, and Africa as discussed in this book (Lyon, 2001).

GIS also allows for advanced analysis and modeling methods to be implemented in support of research efforts. The use of simulations of models provides detail on water movement and transport of materials.

GIS databases and technologies allow for optimization of model results. The running of model scenarios supplies detailed information for the development of plans, management decisions, and informed leadership. Repetitive processing facilitates predictions and forecasting of events.

Certain characteristics of and applications in water resources research lend themselves to GIS databases and GIS analyses. As we know, water flows downhill and supplies a directional characteristic to the modeling efforts. The flow can be determined by gauging, and the simulations of water discharge and water quality above a gauge can be compared with the reality measured at the gauge.

The history of watershed and water resources research has included many models that utilize spatial data. These "traditional" models incorporate data with a spatial basis or the components of spatially averaged data. This is because applications have been dictated by need, and models exhibit sensitivity to these variables.

The future promises a variety of enhanced applications of GIS and allied technologies for water resources and watershed research. We will see better integration of data within databases, and the advent of more uses of three-dimensional visualization of data. Databases will become available in greater numbers and detail, and will be procured through the Internet and the World Wide Web.

The advantages of using GIS for watershed studies has been recognized. It is apparent that the capabilities of GIS are potentially valuable in a number of efforts. Over time the need for information has resulted in the development of appropriate algorithms to facilitate the utility of GIS and statistical or deterministic models. Much traditional modeling and analyses preceded the advent of GIS and many efforts are ongoing to join the approaches and optimize their capabilities. It is these kinds of efforts that have been brought together here to help share the experience of the authors with the readers.

The chapters presented here demonstrate a number of practical applications of theory and applications of GIS technologies to solving real-world problems. Some outstanding authors have contributed. They wish to showcase their high quality efforts, and share their knowledge and experience with other professionals. The result is an informative and practical book that will be enjoyed for years to come.

The book begins with two contributions from Martz and Garbrecht, who are well-known for their thoughtful work in watershed and water resources research. These papers help to characterize the variable that can be obtained through analysis of Digital Elevation Model (DEM) data.

Here, DEM data were preprocessed to identify drainage networks using the Digital Elevation Drainage Network Model (DEDNM). Application of the model and software allows extraction of topographic and topologic drainage network and watershed properties, using available DEM data from 1:24,000 scale USGS quadrangle maps. Data corrections such as removal of depressions and flat areas are addressed and methods discussed. Watershed characteristics are developed such as channel tracing by flow direction using codes to identify links, nodes and subwatersheds. Other computations to develop variables include calculations of slope and overland flow distances, elevations, and associated flow paths.

The second chapter by these two authors further develops the efforts to calculate watershed characteristics. Included are efforts to correct artifacts in DEM data using DEDNM, and use of other algorithms in the TOpographic PArameteriZation program (TOPAZ). The TOPAZ digital analysis system processes and analyzes DEM data to characterize topography, measure parameters, identify surface drainage, subdivide watersheds, and quantify the drainage network.

An additional contribution of this chapter is an evaluation of an automated approach to development of watershed data including segmentation, channel network identification, and subwatershed definition. This is completed for a small watershed in low relief terrain. The channel network generated by DEDNM is compared with a traditional blue-line method to determine the capability and accuracy of the automated approach.

The chapter by Tachikawa, Shiiba, and Takasao presents a TIN-DEM based topographic model which incorporates the advantages of grid-based methods and contour-based methods for generating watershed characteristics. The results were applied to the question of determining the structure of a distributed rainfall-runoff model, and determinations of source of flow and direction of runoff flow.

The chapter by Harvey and Eash provides a history and description of capabilities of the Basinsoft program. This program processes data through modules, and can quantify 27 different basin characteristics. Watershed characteristics can be generated for a given basin or multiple basins, and the repetitive steps can be automated.

The capabilities of Basinsoft were compared to manual measurement in this chapter. Twelve selected drainage basin characteristics were made from USGS topographic data for drainage areas upstream of 11 streamflow gauging stations in Iowa. Results of Wilcoxon signed ranks test indicated that Basinsoft quantifications were not significantly different from manual measurements for 9 of 10 drainage basin characteristics tested. This work was significant because it demonstrated statistically that GIS- and DEM- derived parameters were similar to those derived from traditional, manual measurement procedures. Such validation of automated procedures is a necessary and important step in GIS research and applications.

The chapter by Miller, Guertin and Goodrich presents a methodology for predicting channel shape from watershed characteristics that could be readily derived from commonly available GIS data. A high resolution database was constructed and tested for the Walnut Gulch Experimental Watershed in Arizona. Channel cross-sectional area and width were found to be significantly related to channel order, upstream watershed size, and maximum contributing flow length within a watershed.

The chapter by Mizgalewicz, White, Maidment and Ridd, describes procedures developed to model the water balance of the large-scale Mississippi River flooding event of 1993, and to utilize hydrologic and meteorologic data sets in a GIS database. The goals were to develop a comprehensive understanding of the climate and hydrologic conditions related to the floods. The chapter also presents ideas on how to use visualization software in the depiction of water storage change over the land surface during the flood period.

Starks, Garbrecht, Schiebe, Salisbury, and Waits address the development of hydrographic data layers from DEM. Included in their chapter is the evaluation of reliability of the soil property data extracted from soils coverage and Soil Survey data, and land use and land cover data derived from satellite sensors. The topographic information was developed from TOPAZ, a set of algorithms and software described earlier. The use of Soil Surveys and attribute data was examined, and data employed in the project were taken from the STATSGO database. The concept of critical source area (CSA) was also described and its utility addressed.

The chapter contribution by Thenkabail and Nolte demonstrates the use of remote sensor data, Global Positioning Systems (GPS), and ground data for input to a GIS. In this chapter, three large study areas in Africa were evaluated, and the capabilities of these technologies to locate resources was demonstrated. Fascinating information was developed on the presence of inland valley bottoms, their cultivation, and the proximity of roads to farm fields in a poorly mapped and developed area. The utility of the derived information was great, and the results can demonstrate how these technologies supply details unavailable from any source, and help to fix the location of resources in space and in time.

The chapter by Warner, García-Martinó, Scatena, and Civco is an interesting example of GIS and watershed modeling for use in low-flow watersheds. Low-flow or dry streams in watersheds are difficult to model and this paper demonstrates a thoughtful way to deal with the problems presented by such a watershed in Puerto Rico.

In his chapter, Holman demonstrates how land cover data from Landsat satellites can assist in modeling land cover and use in a rural and urban study area of large size. Numerous and interesting results were generated by evaluation of the database using GIS.

The chapter by Queen, Wold, and Brooks presents a GIS that was developed to manage and analyze landscape data for nine representative subwatersheds in the Nemadji basin of Minnesota. Output from the GIS was used in regression modeling of the relationships between the frequency of soil mass movements and the landscape characteristics. Significant results included the relationship between slump frequency and total forested area and percent nonforested land.

Guan, Moore, and Lostal illustrate in their chapter how a large data set can be accumulated by submitting individual results of projects to a shared database. They discuss the requirements of managing such a database, and making the data available to many users. They also examine the problems associated with changing hardware and software, and how to maintain a useful database over multiple-year time periods.

The chapter by Williams and Lyon presents the utility of historical aerial photos and GIS analyses to inventory and monitor a dynamic water resource over a more than 50-year period. The results address the hydrologic forcing function of fluctuating Great Lakes water levels and their influence over time on coastal wetlands resources. The results have allowed resource managers to examine the natural variability of the wetlands, and any potential influence of human activities.

How the influence of habitat conditions can be evaluated for a given wildlife species is shown in the chapter by Lunetta, Cosentino, Montgomery, Beamer, and Beechie. The issue of salmonid habitat and human activities is a potent one, yet little quantitative, spatial-based work has been done. The combination of GIS analysis and data from a variety of sources including Landsat satellites allowed this difficult issue to be addressed and quantitative information be brought to the dialogue.

In his chapter, Colby examines the role of GIS in pulling together a number of data sources to address watershed characteristics and runoff yields for a rural watershed in Puerto Rico.

The chapter by Cruise and Miller examines the capabilities of remote sensor data to facilitate hydrologic modeling. The combination of remote sensor data, GIS, and models provides the spatial, spectral, and numerical capacity to address topical interests. Results demonstrate that macroscale hydrologic modeling with conventional remote sensing tools can provide simulations of observed climatological features over fairly long time periods.

The chapter by Garbrecht, Martz, and Starks presents the results of preprocessing DEM data to treat problems associated with depressional and flat areas. The additional image processing technologies better capture the channel networks from overland or hillslope areas, and enable automated management of the water flow routing process in large drainages, and help generate network and subcatchment data necessary for rainfall-runoff modeling.

Garofalo demonstrates the practical value of using historical aerial photographs and photointerpretation techniques for historical analyses. This very informative treatment details how valuable information can be derived on hazardous waste disposal areas and the associated water resource characteristics. Use of these techniques can greatly assist users in developing hard-to-obtain factual information on most sites of interest, and for populating GIS databases.

The chapter by Lyon, Ward, Atherton, Senay, and Krill demonstrates a number of techniques that can be applied to agricultural and water resource applications (Ward and Elliot, 1995). The combination of advanced techniques such as remote sensing, GIS, GPS, variable rate applications, detailed soil chemical sampling, and related approaches has greatly improved decision making in agriculture. The use of site-specific or precision agriculture can be done in a cost-friendly manner, and greatly improve production and protect water quality.

BIBLIOGRAPHY

Maidment, D. and D. Djokic, 2000. Hydrologic and Hydraulic Modeling Support with Geographic Information Systems. ESRI Press, Redlands, CA.

Garofalo, D., 2001. Aerial photointerpretation of hazardous waste sites. In Lyon, J., Ed., 2001.

Kennedy, M., 1996. *The Global Positioning System and GIS*. Ann Arbor Press, Chelsea, MI.

Lyon, J., 2001. *Wetland Landscape Characterization*. Ann Arbor Press, Chelsea, MI.

Lyon, J., Ed., 2003. *GIS for Water Resources and Watershed Management*. Taylor & Francis, London.

Van Sickle, J., 2001. *GPS for Land Surveyors*. Ann Arbor Press, Chelsea, MI.

Ward, A. and W. Elliot, 1995. *Environmental Hydrology*. Lewis Publishers, Boca Raton, FL.

CHAPTER 1

Channel Network Delineation and Watershed Segmentation in the TOPAZ Digital Landscape Analysis System

Lawrence W. Martz and Jurgen Garbrecht

INTRODUCTION

Research over the past decade has demonstrated the feasibility of extracting topographic information of hydrological interest directly from digital elevation models (DEM). Techniques are available for extracting slope properties, catchment areas, drainage divides, channel networks and other data (Jenson and Domingue, 1988; Mark, 1988; Moore et al., 1991; Martz and Garbrecht, 1992). These techniques are faster and provide more precise and reproducible measurements than traditional manual techniques applied to topographic maps (Tribe, 1991). As such, they have the potential to greatly assist in the parameterization of hydrologic surface runoff models, especially for larger watersheds (i.e., >10 km^2) where the manual determination of drainage network and subwatershed properties is a tedious, time-consuming, error-prone, and often highly subjective process. The automated techniques also have the advantage of generating digital data that can be readily imported and analyzed by Geographic Information Systems (GIS).

This chapter presents an overview and discussion of the original DEDNM (Digital Elevation Drainage Network Model) computer program developed to measure drainage network and subcatchment parameters directly from a DEM. A detailed discussion of the original program structure and algorithms is available in Martz and Garbrecht (1992). Improvements and modifications to the program continue to be made, although the essential approach to drainage analysis remains largely unchanged. DEDNM now functions as the core component in a more comprehensive digital landscape analysis system known as TOPAZ (TOpographic PArameteriZation) (Garbrecht and Martz, 1999).

DEDNM OVERVIEW

The main purpose of the program is to provide an automated, rapid, and reproducible evaluation of the topographic properties of large watersheds by processing raster DEMs. The primary application target for the program is the parameter determination for hydrologic surface runoff models. The following general objectives guided the program development:

1. Program input should be DEMs that are both commonly available to hydrologists and represent the land surface in sufficient detail for hydrologic analysis.
2. The program should not be limited by the nature of the topography represented in the DEM and should be applicable to both high and low relief terrain.

3. The program should extract all topographic and topologic drainage network and watershed properties relevant to the parameterization of hydrologic surface runoff models.
4. The program should incorporate established and proven algorithms from earlier research, and supplement these with new algorithms as required.
5. The program output should be structured so that it can be used directly for hydrologic model parameterization and be readily imported by a GIS.

The program first preprocesses the DEM to remove depressions and flat areas without modifying other parts of the DEM. Flow simulation algorithms are then applied to determine the catchment area of each grid cell, and to define the boundaries of the watershed to be analyzed. A continuous, unidirectional channel network is delineated by selecting all grid cells with a catchment area in excess of a user-specified critical source area.

Network evaluation involves tracing the course of the channels in the previously defined watershed, and identifying the junctions in the network. Channel tracing is directed by flow direction codes generated in the flow simulation analysis. The channels are traced initially to order the network according to the Strahler system, and to determine the length of each network link and the spatial coordinates of its upstream and downstream end. These data are used, in turn, to calculate the slope and catchment area properties of each link, and the number and average properties of the channel segments of each Strahler order. The channels are traced again to assign unique identification numbers to each node in the network. These node numbers are used to associate link, node, and subwatershed data, and can also be used to optimize flow routing in hydrologic models. A final trace is used to identify the subwatersheds of each source node and of the left and right banks. The individual subwatershed areas are then evaluated to determine their slope and overland flow travel distances.

The program generates both tabular and raster output (Table 1.1). The tabular output gives the watershed and drainage network properties of the total drainage network and of the individual network links. The raster output provides maps of the watershed and drainage network components which can be imported into a GIS, registered to other data layers, and used as templates to extract

TABLE 1.1. Summary of the Major DEDNM Program Outputs

Raster Output	
• DEM, depressions filled	• Template of main watershed
• DEM, depressions filled, relief on flat areas	• Channel network: links by Strahler order
• Extent of depressions and flat areas	• Channel network: links by node number
• Flow direction at each grid cell	• Subwatersheds: by node number and subarea
• Drainage area at each grid cell	code

Tabular Output	
• Main watershed (by Strahler order)	• Individual network links
1. Number of channel segments	1. Strahler order
2. Average channel length	2. Upstream and downstream node coordinates
3. Average channel slope	3. Upstream node number
4. Average upstream and direct drainage area	4. Channel length
• Individual subwatersheds	5. Upstream and downstream node elevation
1. Channel area	6. Upstream and direct drainage area
2. Left-bank, right-bank and source node	
direct drainage area	

additional, hydrologically-relevant information (i.e., soil type, land cover, climate, etc.) for individual subwatersheds and network components.

DEDNM is coded in FORTRAN–77. It is modular to allow easy modification and reorganization to meet specific needs. Each major operation is performed by a set of subroutines called from the main program. Program input is a raster DEM in a standardized format, DEM parameters (i.e., number of rows and columns, elevation range, etc.) and user-specified job parameters that control processing and output. DEDNM keeps all data in dynamic memory during execution to minimize time-consuming I/O operations. While memory requirements are minimized by performing all computations on and storing all elevations as integer data, the processing of large DEMs can create significant memory demands. An option to aggregate the DEM to a coarser grid cell size is provided to permit operation in limited memory or to permit more rapid preliminary analyses of large DEMs. DEDNM can be compiled to run on any hardware platform and, provided sufficient memory is available, can be easily configured to process any size of DEM.

DEM PREPROCESSING

The design objectives for DEDNM emphasize the analysis of generally available DEMs which represent landscapes at a resolution that allows the extraction of hydrologic variables. The 7.5-minute DEMs distributed by the United States Geological Survey (USGS) are typical of such data. They provide elevation values in 1 m increments for 30 m grid cells over areas corresponding to USGS 7.5-minute quadrangles. Much of the United States is now covered by such DEMs, and coverage continues to increase. DEMs with the same general structure and order of resolution are provided by other national jurisdictions and can be generated from SPOT data (Quinn et al., 1991; Tribe, 1991).

A fundamental problem in using DEMs of this order of resolution for hydrologic analysis is the presence of sinks in the data. Sinks are grid cells with no neighbors at a lower elevation and, consequently, with no downslope flow path to a neighbor. By this definition, sinks occur on both flat areas and in closed depressions. Sinks are quite common in DEMs with a spatial and vertical resolution similar to that of the USGS 7.5-minute DEMs. They also tend to be more common in low relief terrain than in high relief terrain. While a few of these sinks may represent real landscape features, the majority are spurious features which arise from interpolation errors during DEM generation, truncation of interpolated values on output, and the limited spatial resolution of the DEM grid (Mark, 1988; Fairchild and Leymarie, 1991; Martz and Garbrecht, 1992).

Figure 1.1 shows the spatial distribution of sinks in a DEM of an 84 km² low relief watershed used as a test data set in developing DEDNM (Garbrecht and Martz, 1997). The DEM elevations are in 0.9 m (3 ft) increments for 30 m grid cells. Ten percent of the watershed is covered by sinks, of which 25% are in closed depressions and 75% are on flat areas. The sinks are concentrated along valley bottoms where local slopes are gentle. The banded appearance of many of the flat areas suggests that they are artifacts of the limited vertical resolution of the DEM.

The ability to simulate flow across sinks is essential for effective hydrologic analysis of DEMs of this type. DEDNM provides this capability through a two-phase operation which has the same effective result as the flow modification method of Jenson and Domingue (1988). The first phase involves filling all closed depressions to the elevation of their local outlet using the method of Martz and de Jong (1988). The subwatershed of a depression is delineated and the lowest outlet on the edge of this area is identified. The elevation of all cells within the subwatershed and below the elevation of the lowest outlet are raised to the outlet elevation to simulate depression filling (Figure 1.2). Complex, nested, or truncated depressions are readily evaluated by the filling algorithm.

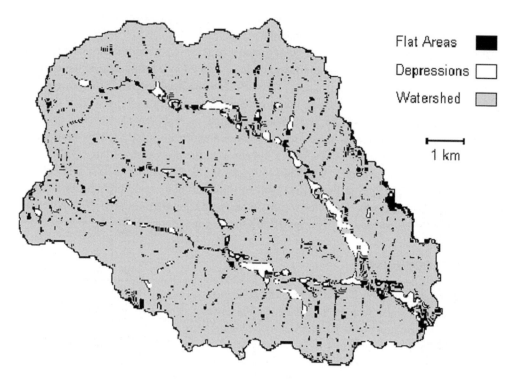

Figure 1.1. Type and spatial extent of sinks in DEM of Bill's Creek watershed in southwestern Oklahoma.

Subtracting elevations in the original DEM from those in the DEM modified by depression-filling gives the location, spatial extent, and depth of depressions.

The second phase imposes relief on all flat areas (both those created by depression-filling and those inherent to the DEM) to permit the unambiguous definition of flow paths across these areas. Two assumptions are implicit to this operation: (1) flat areas are not truly level, but have relief that is simply not detectable at the resolution of the DEM, and (2) this relief directs flow entering or originating on the apparent flat area along a reasonably direct path over the flat area to a point on its perimeter where a downward slope is available. Relief is imposed by adding an elevation increment of 1/1000th of the vertical resolution of the DEM to each cell with no neighbor at a lower elevation (Figure 1.2). This is repeated until no flat areas remain. This approach to the treatment of sinks is superior to the use of prior smoothing to remove sinks because it focuses directly on problem areas without reducing the information content of the DEM elsewhere. It also relies on an assessment of local boundary conditions that bear directly on flow patterns.

The current version of DEDNM incorporated into the TOPAZ digital landscape analysis system introduced two significant improvements for the treatment of depressions and flat areas. These improvements allow the breaching (i.e., lowering) of the outlet of some depressions and apply a two-phase relief imposition algorithm to direct flow across flat areas simultaneously away from higher and toward lower surrounding terrain (Garbrecht and Martz, 1997).

FLOW VECTOR AND DRAINAGE AREA ANALYSIS

Following the preprocessing operation, a flow vector code is assigned to each grid cell using the D8 method (Fairchild and Leymarie, 1991). The flow vector indicates the direction of the

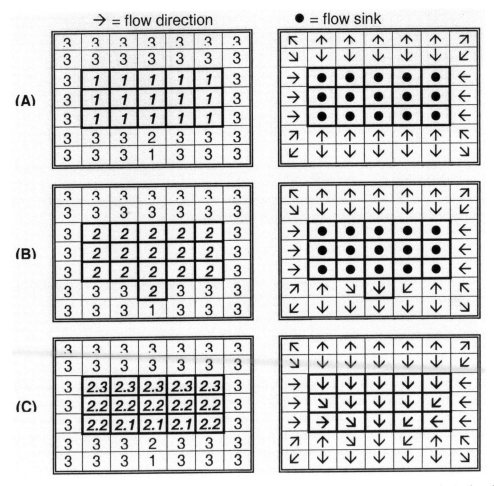

Figure 1.2. DEM preprocessing by DEDNM. Elevations (left column) and associated flow paths (right column) are shown for: (A) original DEM with depression; (B) DEM with flat area created by depression filling; (C) DEM with relief imposed on flat area (for simplicity of illustration, an elevation increment of 1/10th is used).

steepest downward slope to an immediately neighboring cell. Where more than one downward slope maxima exist, the flow vector is arbitrarily assigned to indicate the direction of the maximum first encountered. At cells on the edge of the defined DEM (i.e., cells in the outer rows and columns, or adjacent to a cell with a missing elevation value), the flow vector points away from the defined DEM if no other downward slope to a neighbor is available. All flow originating on or entering a cell is assumed to move in the direction indicated by the flow vector, and no divergent flow out of a cell is accommodated.

The catchment area of each grid cell is determined using the method of Martz and de Jong (1988). The flow vectors are used to follow the path of steepest descent from each cell to the edge of the DEM, and the catchment area of each cell along this path is incremented by one. After a path has been initiated from each cell, the catchment area value accumulated at each cell gives the number of upstream cells which contribute overland flow to that cell (Figure 1.3).

The boundary of the watershed to be analyzed is also determined from the flow vectors. The user specifies the location of the grid cell at watershed outlet, and all grid cells which contribute

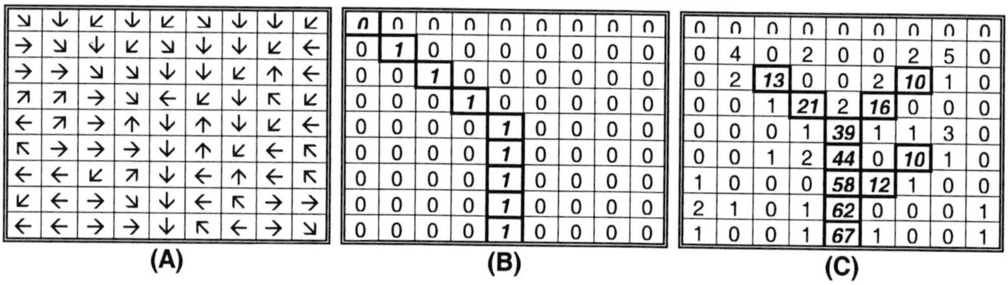

Figure 1.3. Catchment area determination by DEDNM: (A) flow vectors at each DEM cell; (B) catchment area at each DEM cell after the first flow path (highlighted) has been traced from the upper-left cell; (C) catchment area at each DEM cell after flow paths have been traced from all cells. Highlighted cells in (C) are those classified as channels using a critical source area of 10 cells.

overland flow to the outlet cell are identified. This provides a mask of the watershed that is used for subsequent operations.

CHANNEL NETWORK ANALYSIS

The channel network within the watershed is delineated from the catchment area grid. All cells with a catchment area greater than a user-specified critical source area are classified as part of the channel network. This yields a fully connected, unidirectional network. However, the network may contain some very short exterior links which represent valley side indentations, gully outlets, and other features that normally would not be classified as part of the channel network. The network is pruned to remove these spurious exterior links. The pruning involves tracing each exterior link downstream from its source to determine if a user-specified threshold distance (the minimum source channel length) can be traveled before a junction is reached. If so, the link is retained. If not, the link is a candidate for removal. Where two links less than the threshold length join, the shorter link is removed and the remaining link is reevaluated to determine if it now meets the threshold length criterion for retention. A slightly more involved analysis is required at junctions of more than two links.

Once the network has been pruned, the length, starting and ending cell coordinates, and Strahler order of each channel link are determined. Each exterior link is traced downstream from its source until a junction is reached. Each cell along the link upstream of the junction is marked as Strahler order 1, while the cell at the junction is marked as Strahler order 2. On reaching a junction, the link coordinates and link length are stored, and the trace terminates. Once all exterior links have been traced, interior links of successively higher Strahler order are evaluated.

Each junction where a cell was previously marked as being of the current Strahler order (initially order 2) is examined to determine if all links entering the junction have already been assigned a lower order. If this is the case, the junction is the upstream end of a channel segment of the current order, and a trace is initiated. The trace proceeds downstream from the junction, with each cell along the interior link being assigned the current Strahler order until another junction is reached. On reaching a junction, the length and the upstream and downstream coordinates of the link are stored, and the other links entering the junction are examined. If all other links entering the junction were previously assigned a lower order, the order of the downstream link will not increase and the trace continues. If this is not the case, the junction is flagged as being of the next

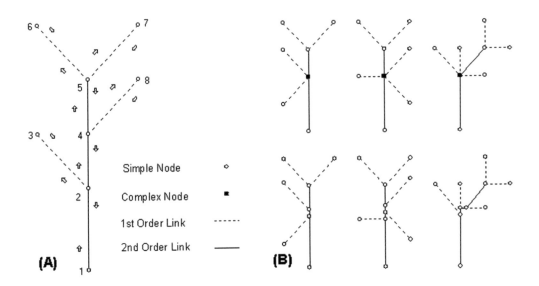

Figure 1.4. Node numbering by DEDNM: (A) the general method of assigning node numbers by tracing the network as shown by arrows; (B) the upper row shows networks with complex nodes (i.e., junctions of more than two channels) in the grid representation of the channel system and the lower row shows corresponding networks in which the complex junctions are effectively replaced by multiple simple nodes in the DEDNM node numbering procedure. The multiple simple nodes are assumed to fall within the grid cell containing the complex node which they replace.

highest order, and the trace terminates. Analysis continues until all links in the network have been ordered.

Following network ordering, the link coordinates are used to find the elevation and the catchment area at the upstream and downstream end of each link. These are used to calculate the longitudinal slope and direct drainage area, respectively, for each link in the network. In addition, the number of channel segments, and the average slope, length, and total and direct drainage area of the segments of each Strahler order are calculated.

A unique identification number is then assigned to each network node using the system described by Garbrecht (1988). Flow vectors are used to simulate a walk along the left bank of the channel beginning and ending at the watershed outlet. Each node is assigned a number the first time it is passed during the walk (Figure 1.4). These numbers can be used by hydrologic models to optimize flow routing, and are used by DEDNM to associate node, link, and subwatershed data. Channel networks often have some junctions where more than two links join. This situation is treated by assuming that more than one node occupies a single grid cell, and assigning one number to each assumed node. Some special analysis is required to ensure that the node number assignment remains consistent with the network ordering applied earlier. The method by which this is done is beyond the scope of this chapter, but some illustrative results are presented in Figure 1.4.

SUBWATERSHED DELINEATION AND ANALYSIS

Subwatershed delineation is accomplished in two stages. In the first stage, flow vectors are used to follow each link from its upstream end to its downstream end, and to identify the cells

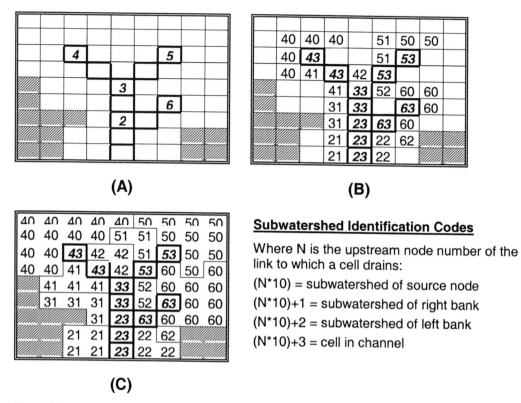

(A)   **(B)**

(C)

Subwatershed Identification Codes

Where N is the upstream node number of the link to which a cell drains:

(N*10) = subwatershed of source node

(N*10)+1 = subwatershed of right bank

(N*10)+2 = subwatershed of left bank

(N*10)+3 = cell in channel

Figure 1.5. Subwatersheds defined by DEDNM from the flow vectors in Figure 1.3: (A) highlighted channel network cells and node numbers of sources and junctions; (B) subwatershed codes assigned to all cells adjacent to the network cells; (C) subwatershed codes assigned to all cells in the watershed. Shaded cells are outside the watershed.

which are immediately adjacent and contribute flow to the link. These cells are classified according to whether they are on the left bank, the right bank or, in the case of exterior links, contribute flow directly to a source node. A subwatershed identification code based on the previously assigned node numbers is applied to each classified cell. In the second stage, the entire grid is scanned and all cells in the main watershed that have not been assigned a subwatershed code are evaluated to determine if they contribute flow to a cell which has been assigned a subwatershed code. If so, it is assigned the code of the cell to which it flows. This is repeated until all cells have been assigned a code. The subwatershed identification process is illustrated in Figure 1.5.

Once the subwatersheds have been identified, they can be evaluated to determine their individual slope, aspect, and overland flow distance properties. These variables were measured directly in the original DEDNM. In the TOPAZ digital landscape analysis system, of which DEDNM is now a part, a separate raster processing program (RASPRO) now performs this and several other functions (Garbrecht and Martz, 1997).

CONCLUSIONS

The computer program DEDNM was developed with the primary objective to assist the rapid parameterization of hydrologic surface runoff models using DEMs similar to those provided by the USGS for 7.5-minute quadrangles. The set of algorithms developed for this purpose provide:

1. Raster maps including the corrected DEM, the flow patterns, the drainage network, and the extent of the main watershed and its subwatersheds.
2. Channel and subwatershed indexing by node identification numbers.
3. Tables of channel and subwatershed attributes, and of drainage network and subwatershed composition.

DEDNM is able to process limited resolution DEM data of low and high relief terrain to provide results that are both conceptually valid and consistent with the topographic characteristics of the landscapes to which it is applied. A more comprehensive evaluation of the capabilities of the program to reproduce the results of a traditional manual analysis of the blue-line network of topographic maps has been undertaken and is reported in the following chapter.

While the essential approach to drainage analysis introduced in DEDNM has remained largely unchanged, the software has been continuously modified and improved. One of the more significant changes has been the incorporation of DEDNM into TOPAZ, a more comprehensive digital landscape analysis system (Garbrecht and Martz, 1997), in which DEDNM provides the core function of network delineation and watershed segmentation from raster digital elevation data.

The TOPAZ digital landscape analysis system processes and evaluates raster DEMs to identify topographic features, measure topographic parameters, define surface drainage, subdivide watersheds along drainage divides, quantify the drainage network and parameterize subcatchments. The core drainage analysis provided by DEDNM allows TOPAZ to maintain consistency between all derived data, the initial input topography, and the physics underlying energy and water flux processes at the landscape surface. The primary objective of TOPAZ is to support hydrologic modeling and analysis, but it can also be used to address a variety of geomorphological, environmental and remote sensing applications.

REFERENCES

Fairchild, J., and P. Leymarie, 1991. Drainage networks from grid digital elevation models. *Water Resources Research*, 27(4):29–61.

Garbrecht, J., 1988. Determination of the execution sequence of channel flow for cascade routing in a drainage network. *Hydrosoft* 1(3):129–138.

Garbrecht, J., and L.W Martz, 1999. *TOPAZ: An Automated Digital Landscape Analysis Tool for Topographic Evaluation, Drainage Identification, Watershed Segmentation and Subcatchment Parameterization; TOPAZ Overview*. USDA-ARS Publication No. GRL 99–1, 26 pp.

Garbrecht, J., and L.W. Martz, 1997. The assignment of drainage direction over flat surfaces in raster digital elevation models. *Journal of Hydrology*, 193: 204–213.

Jenson, S.K., and J.O. Domingue, 1988. Extracting topographic structure from digital elevation data for geographical information system analysis. *Photogrametric Engineering and Remote Sensing*, 54(11):1593–1600.

Mark, D.M., 1988. Network models in geomorphology. In *Modeling Geomorphological Systems*, M.G. Anderson, Ed., John Wiley and Sons, pp. 73–96.

Martz, L.W., and E. de Jong, 1988. CATCH: A FORTRAN program for measuring catchment area from digital elevation models. *Computers and Geosciences*, 14(5):627–640.

Martz, L.W., and J. Garbrecht, 1992. Numerical definition of drainage networks and subcatchment areas from digital elevation models. *Computers and Geosciences*, 18(6):747–761.

Martz, L.W., and J. Garbrecht, 1998. The treatment of flat areas and closed depressions in automated drainage analysis of raster digital elevation models. *Hydrological Processes*, 12: 843–855.

Moore, I.D., R.B. Grayson, and A.R. Ladson, 1991. Digital terrain modeling: A review of hydro-logical, geomorphological and biological applications. *Hydrological Processes,* 5(1):3–30.

Quinn, P., K. Beven, P. Chevallier, and O. Planchon, 1991. The prediction of hillslope flow paths for distributed hydrological modelling using digital terrain models. *Hydrological Processes,* 5(1):59–79.

Tribe, A., 1991. Automated recognition of valley heads from digital elevation models. *Earth Surface Processes and Landforms,* 16(1):33–49.

CHAPTER 2

Assessing the Performance of Automated Watershed Segmentation from Digital Elevation Models

By Jurgen Garbrecht and Lawrence W. Martz

INTRODUCTION

Watershed segmentation and channel network definition is often required in distributed hydrologic modeling. Manual segmentation from maps is a tedious, time-consuming, and subjective task, particularly for large watersheds. The automated watershed segmentation and extraction of channel network and subwatershed properties from raster elevation data represents a convenient and rapid way to parameterize a watershed. The increasing availability of DEM coverage for many areas of the United States makes this automated watershed segmentation and characterization a promising approach for a wide range of hydrologic investigations. However, assessment of the generated watershed data beyond the usual visual inspection is required to develop confidence in the automated approach.

Early research on automated landscape analysis focused on algorithm development and treatment of unique situations, such as depressions, flat areas, or the connectivity of the network. DEM processing models which identify upward concave areas (Pueker and Douglas, 1975; Jenson, 1985; Band, 1986) often produce discontinuous network segments that must subsequently be connected (O'Callaghan and Mark, 1984) and may require additional adjustments to produce a reasonable pattern (Douglas, 1986). Other models rely on flow routing concepts. In this approach, the steepest downslope direction defines the flow paths (Jenson and Domingue, 1988; Martz and de Jong, 1988; Morris and Heerdegen, 1988). For either approach, problems arise in low relief terrain when the vertical resolution of the DEM is insufficient to identify either upward concave areas or a downslope flow direction. Little work has been done to extend DEM processing methods to low relief terrain, such as found in the central plains of the U.S.

This chapter investigates the performance of automated watershed segmentation, channel network identification, and subwatershed definition in low relief terrain for an 84 km^2 watershed. The landscape analysis computer program chosen from this study is the Digital Elevation Drainage Network Model DEDNM (Martz and Garbrecht, 1992, 2001). Program DEDNM is the main component of a larger landscape analysis program called TOPAZ (TOpographic PArameteriZation) (Garbrecht and Martz, 1999, 2000). The approach used by program DEDNM is similar to that of other DEM processing models that are based on flow routing concepts, but it includes enhancements for processing low relief landscapes where the rate of elevation change may be only a few meters per kilometer over large areas. The program is designed for the resolution of the USGS 7.5′ DEMs (1 meter or 1 foot elevation increments for 30 m grid cells) and targets parameters that are of hydrologic interest. Even though program DEDNM was designed primarily for hydrologic and water resources investigations, it is equally applicable to address a variety of geomorphological, environmental, and remote sensing applications. The performance of program DEDNM is evaluated by comparing the generated watershed subdivisions, channel network, and other derived parameters to those obtained by traditional evaluation methods using USGS 7.5′ minute topographic maps.

THE APPLICATION WATERSHED

Bills Creek watershed drains an 84 km^2 area of the Little Washita River watershed, a USDA-ARS experimental watershed in southwestern Oklahoma. The terrain consists of gently rolling hills, and land use is predominantly rangeland with some cultivated areas. The source and the outlet of Bills Creek are at 420 m and 336 m above MSL, respectively. With a longitudinal distance of 16.5 km, the average main channel slope is about 0.005 m/m. Channel slopes as low as 0.003 m/m and flat flood plain areas extending over several hectares are not uncommon toward the outlet of Bills Creek watershed.

USGS 7.5′ DEM coverages from the geographic area under consideration are not available, and a previously generated DEM of the ARS experimental watershed is used to conduct this study. The DEM was generated in 1987 by the NASA Stennis Laboratory, Slidell, Louisiana, from USGS 7.5′ topographic maps. Elevation data are given for 30 m grid cells in 0.91 m (3-foot) increments. These DEM characteristics are similar to the USGS 7.5′ DEM coverages which have a 30 m grid spacing and a 1 m vertical resolution. A three-dimensional, vertically enhanced representation of Bills Creek watershed is shown in Figure 2.1 (see color section). With a vertical resolution of 0.91 m, parts of the gently sloping floodplains appear as flat areas and include a number of apparent depressions or pits which are artifacts of the DEM. Such spurious depressions are often encountered in low resolution DEMs of terrain with limited relief and usually arise from input errors, interpolation procedures, elevation rounding and limited DEM resolution. These produce under- and overestimation of elevation values of individual or groups of DEM cells which then can result in spurious depressions. The extent of the flat areas and depressions in Bills Creek watershed are illustrated in Figure 2.2 (see color section). They can be seen to be closely associated with channel bottom lands and drainage divides. The low channel slopes, flat areas, and spurious depressions in the DEM of Bills Creek watershed make it an appropriate low relief application for evaluating watershed segmentation and channel network generation capabilities of program DEDNM.

The channel network, defined by the blue-line channels on the USGS 7.5′ topographic maps (Figure 2.3a), is the map-based network against which the generated network is assessed. It should be recognized that a channel network is a dynamic drainage feature that can change in time as a function of climate, land use, and other land surface parameters. Thus, any channel network defined from topography alone is the reflection of past runoff and erosional activities and may not necessarily be representative of current runoff conditions. Other methods such as contour crenulation or slope analysis can also be used to define the channel network, but they also have limitations. The blue-line method (Morisawa, 1957) was selected because it is the simplest and one that most readers are familiar with, and despite its limitations, it is a standard that is readily available for most watersheds in the United States. The map-based network of Bills Creek watershed is of 5th Strahler order (Strahler, 1957) and of magnitude 180 following Shreve's (Shreve, 1967) classification.

BOUNDARY CONDITIONS FOR WATERSHED SEGMENTATION

Program DEDNM requires that two channel network parameters be specified for the automated watershed segmentation: the critical source area and the minimum source channel length. The crit-

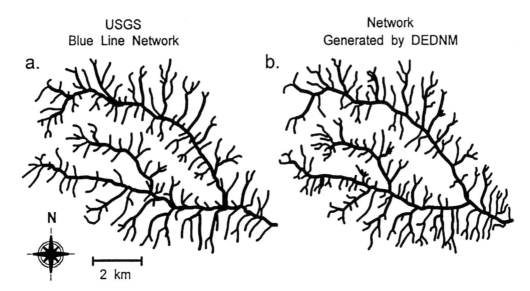

Figures 2.3a and 2.3b. Channel network of Bills Creek watershed: (a) blue-line network from USGS 7.5′ topographic maps; (b) generated network by DEDNM.

ical source area is the drainage area required to support a permanent channel. Its value is related to soil characteristics, vegetation cover, climatic conditions, and terrain slope. It also varies with map scale when maps are used as the basis for network definition (Scheidegger, 1966). For the present application, the critical source area was determined from the USGS 7.5′ topographic maps and was about 9 ha. The minimum channel length for source channels is a parameter that is necessary to control the identification of very short channels that satisfy the critical source area criterion, but have no real significance at the scale of the USGS 7.5′ topographic maps. Therefore, a threshold length below which first order channels are not generated is specified. This threshold length for Bills Creek was determined from the USGS topographic maps to be 130 m. With only these two input channel network parameters, the DEM can be processed by program DEDNM, and a segmented watershed, and channel network and subcatchment parameters can be extracted from the DEM. As part of this processing, the spurious depressions in the DEM are first removed by raising the elevation of the cells within the depressions to the elevation of the lowest outlet cell on the outside edge of the depression. The watershed segmentation and channel network and subcatchment parameter extraction is then performed on this depression-free DEM.

RESULTS

Program DEDNM generates raster maps of the channel network, subwatersheds, and other drainage parameters. The watershed segmentation is represented by the drainage boundaries of the subcatchments and the channel network. As examples of generated raster maps, the raster map of elevation contours, flow vectors, and subwatersheds boundaries, including the generated channel network, are displayed in Figures 2.4, 2.5, and 2.6, respectively (see color section). Program DEDNM also produces attribute tables that can be used for subwatershed parameterization, distributed surface runoff modeling, or other purposes. In the following discussion, network and sub-

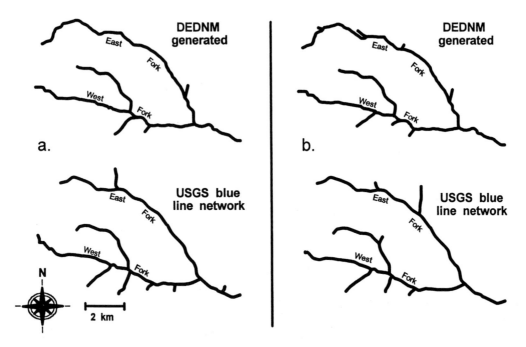

Figures 2.7a and 2.7 b. Generated and measured channel networks for (a) channels of third and higher Strahler order; (b) channels of fifth and higher Shreve magnitude.

watershed raster and attribute data are compared to values derived from USGS 7.5′ topographic maps by traditional methods.

Evaluation by Visual Appearance

The channel network generated by program DEDNM (Figure 2.3b) is similar to the blue-line map-based network derived from the USGS 7.5′ topographic maps (Figure 2.3a). Differences are apparent, particularly at the level of source channels. Some source channels are generated at locations where there are none on the USGS maps, although contour crenulation suggests the presence of a channel in most of these cases. The reverse is also true; source channels with small upstream source area are shown on the USGS maps, but are not generated by program DEDNM. Other source channels differ by their length. The primary reasons for these differences are the ambiguity in the definition of a source of a channel, natural variation in channel characteristics within the network, and the use of only two parameters to generate the entire network. The two-parameter approach used by program DEDNM (critical source area and minimum source channel length) can reproduce average channel properties, but cannot account for random or spatial variability of these properties within the watershed. When such spatial variability is important, the user must either apply the program to each homogeneous subarea separately, or introduce a variable channel maintenance constant and minimum source channel length.

The differences between the two networks diminish for higher channel magnitude or Strahler order. This increasing similarity is demonstrated for third and higher Strahler order channels (Figure 2.7a), and for fifth and higher Shreve magnitude channels (Figure 2.7b). In both cases, the main channel, the East and West Fork, and several larger tributaries are well reproduced. The few remaining differences are channels that exist in both networks, but are not represented because they did not meet the order or magnitude selection criterium used to generate the illustrations.

Comparison of Selected Channel Network Parameters

The selected parameters used to compare the generated and map-based channel networks are listed in Table 2.1 and represent the topologic, geometric, and hydrographic characteristics of the networks. The parameter values are representative of the average network characteristics of the entire network and do not reflect spatial variations within the watershed. As indicated earlier, spatial variability can be accounted for by either evaluating each homogeneous subarea separately, or introducing a variable channel maintenance constant and minimum source channel length.

The watershed Strahler order and the Shreve magnitude are nearly identical for the generated and map-based channel networks. The number of channel links and the link lengths are reproduced with a mean difference of less than 2%.

The channel slope values shown in Table 2.1 are length weighted values; i.e., long channels are assumed to be more representative of average network conditions. The mean slope of the generated channel links is about 23% lower than the slope derived from the map-based network. For interior links the difference is 10%, whereas for exterior links it is 27%, suggesting that exterior links are the primary reason for the discrepancy. Exterior links are the same as first order channels or source channels.

Table 2.1. Selected Network Parameters for Generated and Map-Based Networks and Corresponding Deviations

Parameter	Generated Network	Map-Based Network	Deviation %
Strahler order	5	5	0.0
Shreve magnitude	182	180	1.1
Number of channel links total	363	359	1.1
Exterior links	182	180	1.1
Interior links	181	179	1.1
Channel length [m]			
Total length	188706	187980	0.4
Mean link	520	524	−0.8
Mean exterior link	648	643	0.8
Mean interior link	391	407	−3.9
Channel slope [m/m]			
Mean	0.0170	0.0221	−23.1
Exterior links	0.0212	0.0289	−26.6
Interior links	0.0101	0.0112	−9.8
East Bills Creek	0.0055	0.0051	7.8
West Bills Creek	0.0066	0.0063	4.7
Drainage area [ha]			
Total area	8405	8442	−0.4
Mean links	23.2	23.7	−2.1
Above channel source	9.4	9.0	4.4
Drainage density [1/m]	0.0023	0.0022	4.5

The ambiguous and subjective definition of source channels, natural variations in source chan-nel characteristics, the coarse vertical resolution of the DEM, and the use of only two parameters to generate the entire network are believed to lead to the observed differences. A more representa-tive comparison of channel slope is achieved by computing the slope over long channel stretches, such as for the East and West Fork of Bills Creek. For these two channel stretches, the mean slope from source to watershed outlet is reproduced with less than 10% discrepancy. This illustrates that even though large differences in slope may exist for individual channel links, the average slope over long channel stretches is reasonably approximated by program DEDNM.

The total watershed area and mean direct drainage area feeding channel sources and channel links are reproduced within 4%. Finally, the drainage density of the generated network is within 5% of that of the validation network.

Comparison of Channel Network Composition

The channel network composition is quantified by the bifurcation, length, slope, and upstream area ratios, as defined by Horton (1945) and Schumm (1956). These ratios define the rate of change of a variable with channel order and incorporate a relative measure of the variable magni-tude for each order. The channel network composition analysis complements the previous analysis of selected network parameters because it measures change of parameter value with channel order.

The average value of the four ratios for the generated network is within 4% of that obtained for the map-based network (Table 2.2). The slope and relative position of the regression lines in Fig-ures 2.8a and 2.8b graphically represent the ratios. In all four cases, the regression lines are quite similar in slope and position. This close agreement shows that the general character of a system of channels is reproduced by the automated network extraction. Variations in parameter values for in-dividual Strahler orders are primarily the result of the stochastic nature of the Strahler ordering sys-tem. As previously reported by Gregory and Walling (1973), and Scheidegger (1966 and 1970), the stochastic aspect of Strahler's ordering sometimes result in different order values for corresponding channel segments in two very similar networks. This is why direct comparisons of parameter values for individual Strahler order are not necessarily conclusive and are not performed here.

Table 2.2. Ratios of Channel Network Composition

Parameter	Generated Network	Validation Network	Deviation %
Bifurcation ratio	4.00	3.92	2.0
Length ratio	1.87	2.02	−7.4
Slope ratio	1.79	1.90	5.8
Area ratio	4.54	4.54	0.0

DISCUSSION AND CONCLUSIONS

In this chapter, the channel network generated by program DEDNM for an 84 km^2 low relief watershed is compared to the one defined by the blue-line method on the USGS 7.5' topographic maps. The selected watershed included flat areas and depressions along the valley bottom, and flat areas near drainage divides. The depressions are usually artifacts of the DEM and have been re-moved by raising the elevation of the cells within the depression to the elevation of the lowest out-let cell on the outside edge of the depression. The watershed segmentation and channel network and subcatchment properties are generated from this depression-free DEM.

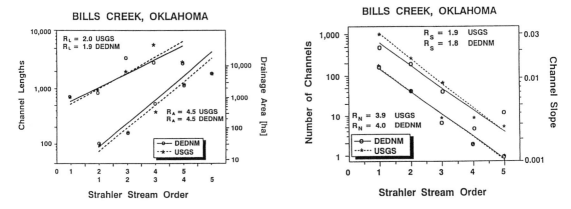

Figures 2.8a and 2.8b. Channel network composition: (a) length and area ratios, R_L and R_A; (b) bifurcation and slope ratios, R_N and R_S.

The visual appearances of the generated and map-based channel networks are very similar, and so are the channel network composition parameters which display an average discrepancy of less than 5%. Other selected network parameters that describe the channel network also show good correspondence. For example, the generated watershed Strahler order, the Shreve magnitude, the number of channel links, the channel link length, the drainage area, and the drainage density are within less than 5% of those of the map-based channel network. The average discrepancy for all parameters used in this study is also less than 5%. The largest differences are found for channel slope. The reason for the larger discrepancies in longitudinal channel slope values is primarily the ambiguous definition of first order channels in the map-based channel network, the coarse DEM resolution, and natural variations in channel characteristics within the network which cannot be reproduced by the two-parameter model of program DEDNM. Spatial differences between generated and map-based channel networks occur because channel generation criteria are applied uniformly to the entire network and, therefore, cannot reproduce spatial variability within the channel network. For investigations where spatial variation is important, program DEDNM should be applied to subareas that are homogeneous. In general, the close agreement between the various parameters describing the overall channel network and subwatershed characteristics demonstrates the ability of program DEDNM to overcome the problems associated with ill-defined drainage boundaries and indeterminate flow paths in low relief terrain.

Based on the experience of this application, further improvements have been introduced into program DEDNM. These include the capability to reproduce spatial variability in the generated channel network and subcatchments within the watershed and to treat spurious depressions by a combination of breaching and filling which is more appropriate given that spurious depressions arise from elevation over- and underestimation. These improvements are covered in a separate chapter of this book. Program DEDNM with these improvements and other additions are available in a software package called TOPAZ (TOpographic PArameteriZation) (Garbrecht and Martz, 1997).

REFERENCES

Band, L. E., 1986. Topographic partition of watersheds with digital elevation models. *Water Resources Research*, 22(1):15–24.

Douglas, D. H., 1986. Experiments to locate ridges and channels to create a new type of Digital Elevation Model. *Cartographica* 23(4):29–61.

Garbrecht, J., and L. W. Martz, 1999. TOPAZ: An Automated Digital Landscape Analysis Tool for Topographic Evaluation, Drainage Identification, Watershed Segmentation and Subcatchment Parameterization; TOPAZ Overview. U.S. Department of Agriculture, Agricultural Research Service, Grazinglands Research Laboratory, El Reno, OK, ARS Publication No. GRL 99–1, 26 pp.

Garbrecht, J., and L. W. Martz, 2000. TOPAZ: An Automated Digital Landscape Analytical Tool for Topographic Evaluation, Drainage Identification, Watershed Segmentation and Subcatchment Parameterization: TOPAZ User Manual. U.S. Department of Agriculture, Agricultural Research Service, Grazinglands Research Laboratory, El Reno, OK, ARS Pub. No. GRL 2-00, 144 pp.

Gregory, K. J., and D. E. Walling, 1973. *Drainage Basin Form and Process: A Geomorphologic Approach.* John Wiley and Sons, New York, NY.

Horton, R. E., 1945. Erosional development of streams and their drainage basins; Hydrophysical approach to quantitative morphology. *Bul. Geol. Soc. Am.,* 56:275–370.

Jenson, S. K., 1985. Automated Derivation of Hydrologic Basin Characteristics from Digital Elevation Data. *Proc. Auto-Carto* 7, Digital Repres. of Spatial Knowledge, Washington, D.C., pp. 301–310.

Jenson, S. K., and J. Q. Domingue, 1988. Extracting topographic structures from digital elevation data for Geographic Information System analysis. *Photogrammetric Engineering and Remote Sensing,* 54(11):1593–1600.

Martz, L. W., and E. DeJong, 1988. CATCH: A FORTRAN program for measuring catchment area from Digital Elevation Models. *Computers and Geosciences,* 14(5):627–640.

Martz, L. W., and J. Garbrecht, 1992. Numerical definition of drainage network and subcatchment areas from Digital Elevation Models. *Computers and Geosciences,* 18(6):747–761.

Martz, L. W., and J. Garbrecht, 2001. Channel network delineation and watershed segmentation in the TOPAZ digital landscape analysis system. In Lyon, J., Ed., 2001, *GIS for Water Resources and Watershed Management.* Ann Arbor Press, Chelsea, MI.

Morisawa, M., 1957. Accuracy of determination of stream lengths from topographic maps. *Trans. Am. Geo. Un.,* 38(1):86–88.

Morris, D. G., and R. G. Heerdegen, 1988. Automatically derived catchment boundary and channel networks and their hydrological applications. *Geomorphology,* 1(2):131–141.

O'Callaghan, J. F., and D. M. Mark, 1984. The extraction of drainage networks from digital elevation data. *Computer Vision Graphics and Image Processing,* 28:323–344.

Peucker, T. K., and D. H. Douglas, 1975. Detection of surface specific points by local parallel processing of discrete terrain elevation data. *Computer Vision Graphics and Image Processing,* 4(3):375–387.

Scheidegger, A. E., 1966. Effect of map scale on stream orders. *Bul. Int. Assoc. Sci. Hydrol.,* XI, 3.

Scheidegger, A. E., 1970. *Theoretical Geomorphology.* Second Revised Edition, Springer Verlag, New York, NY.

Shreve, R. L., 1967. Infinite topologically random channel networks. *J. Geol.,* 75:178–186.

Schumm, S. A., 1956. Evolution of drainage systems and slopes in badlands at Perth Amboy, New Jersey. *Bul. Geol. Soc. Am.,* 67:595–646.

Strahler, A. N., 1957. Quantitative analysis of watershed geomorphology. *Trans. Am. Geophys. Union,* 38(6):913–920.

CHAPTER 3

Development of a Basin Geomorphic Information System Using a TIN-DEM Data Structure

Yasuto Tachikawa, Michiharu Shiiba, and Takuma Takasao

INTRODUCTION

When developing a distributed rainfall-runoff model using Digital Elevation Models, it is important to consider the method with which a spatial distribution of elevations is represented, because the method of a surface representation determines the structure of a distributed rainfall-runoff model. Three principal methods for structuring a network of elevation data are square-grid networks, contour-based networks, and triangulated irregular networks (Moore et al., 1991).

Using contour-based networks, a watershed basin can be subdivided into irregular polygons bounded by contour lines and adjacent to their orthogonals (flow trajectories) that define the boundaries of drainage areas (O'Loughlin, 1986; Moore et al., 1988). Moore and Foster (1990) modified these methods and provided a structure for modeling overland flow, TAPES. For dynamic hydrologic modeling, contour-based methods have a great advantage in considering the directions of water flow, but they need heavy data storage and much computation time.

Square-grid networks are the most common form of DEMs used for topographic analysis of a river basin (O'Callaghan and Mark, 1984; Band, 1986; Hutchinson, 1989; Tarboton et al., 1989; Takasao and Takara, 1989; Takara and Takasao, 1991), and rainfall runoff modeling (Lu et al., 1989; Wyss et al., 1990; Quinn et al., 1991). Grid-based DEMs have advantages for their ease of computational implementation, efficiency, and availability of topographic databases. However, when considering the directions of water flow, these methods are not appropriate for hydrological applications because those trajectories represent only crudely the movements of water from one grid to one of the eight neighboring grids.

A more applicable approach for hydrological modeling is the Triangulated Irregular Networks (TINs). Palacios and Curvas (1986; 1991) made it possible to delineate river-course and ridge of a watershed basin automatically with these methods and to simulate surface runoff production. Jett et al.(1979), Jones et al.(1990) and Vieux (1991) also used TINs for representation of a watershed basin.

This chapter describes a TIN-based topographic model which incorporates the advantages of grid-based methods and contour-based methods. First, a topographic surface is represented by a TIN-DEM generated by a grid-DEM and a DLG (Digital Line Graph) of river courses. Then, these triangle facets are subdivided by the steepest ascent lines (flow trajectories), so each triangle has only one side through which water flows out. Using these triangles, the discretization of a basin similar to contour-based methods is realized.

TIN-DEMS DATA STRUCTURE

In the TIN-DEMs generating system for representing a natural topography of a basin, three datasets are produced: (1) a triangle network data set, (2) a vertex data set, and (3) a channel network data set. A sample triangle data set and its network are illustrated in Tables 3.1, 3.2, and Figure 3.1. Each of the triangles, squares, and vertices is indexed by a number which is given to specify it.

The vertex data set contains the x, y, and z values of the vertices. The triangle network data set contains the properties of triangles. Each triangle is described by an index of the square in which the triangle is contained; indices of its three vertices; indices of three triangles which are adjacent to the triangle; three 'side-attribute-indices' which specify whether water flows into the side, along the side, or out of the side; three 'side-component-indices' which specify whether the side forms a part of valley, channel, slope, ridge, or boundary of a study area; and unit normal vectors of a triangular facet. The indices of vertices, the side-attribute-indices, and the side-component-indices are ordered in a counterclockwise direction.

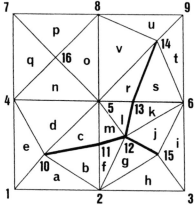

Figure 3.1. Sample triangle network.

Table 3.1. Triangle Network Data Set for Sample Triangle Network Shown in Figure 3.1

Triangle	No. of Squares	No. of Vertices			Adjacent Triangles			Side-Attribute Indices[a]			Side-Component Indices[b]			Unit Normal Vectors		
a	1	1	2	10	null	b	e	3	1	2	0	2	2	−0.71	0.71	0.07
b	1	2	11	10	f	c	a	2	1	3	2	3	2	−0.71	0.71	0.07
c	1	11	5	10	m	d	b	3	3	1	2	2	3	−0.89	−0.41	0.09
.	.	.			.			.			.			.		
.	.	.			.			.			.			.		
.	.	.			.			.			.			.		

[a]Side-Attribute Index: 1 = out-flow side; 2 = along-flow side; 3 = in-flow side.
[b]Side-Component Index: 0 = boundary of TIN-DEM; 1 = valley segment; 2 = slope segment; 3 = channel segment; 4 = ridge segment.

A side-attribute index of a side is set to be 1, 2, or 3, depending on whether water flows out of a side, along a side, or into a side, and the side is defined as an out-flow-side, an along-flow-side, or an in-flow-side, respectively. Whether water flows out of a side, along a side, or into a side is decided by the cross product of the steepest descent vector of a triangle and a side of the triangle.

Table 3.2. Vertex Data Set

Vertex	x	y	z
1	25.00	100.00	301.25
2	50.00	100.00	287.55
3	75.00	100.00	288.89
.	.	.	.
.	.	.	.
.	.	.	.

For example, in Figure 3.2 water on triangle abc flows out to the adjacent triangle through ab. In this case, the z-component of the cross product g x ab is positive, so the side-attribute-index of ab is set to be 1. Similarly, water flows into this triangle from the adjacent triangles through bc and ca. In this case, the z-components of g x bc and g x ca are negative, so side-attribute-indices of bc and ca are set to be 3. If water flows along a side, the cross product is equal to zero, and the side-attribute-index is set to be 2.

A side-component-index of a side is set to be 0, 1, 2, 3, or

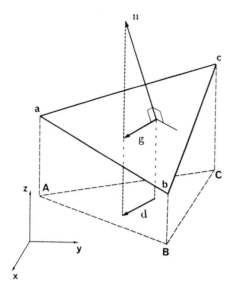

Figure 3.2. Sample triangle facet.

4, depending on whether the side constitutes a part of a boundary of a TIN-DEM, valley, slope, channel, or ridge, respectively. What value a side-component-index is set to be is decided by the side-attribute-indices of the sides which are held in common by the adjacent triangles.

If the common sides of the adjacent triangles are composed of an out-flow-side and an out-flow-side, the sides represent part of a valley. Similarly, if composed of an in-flow-side and an in-flow-side, the sides represent part of a ridge. The relation between a side-attribute-index and a side-component-index is shown in Table 3.3.

A sample channel data set and its network is illustrated in Table 3.4 and Figure 3.3. For a logical representation of a channel network in a computer, a channel network is represented by a set of links which are composed of the sections of a channel network between the terminal point of a channel network and a confluence, a confluence and another confluence, or a confluence and the upstream ends. Each link is also indexed by a number which is given to specify it. The channel network data set is represented by the index of a link, the index of the downstream link, the indices of the upstream links, the indices of vertices which form the link, and the indices of the triangles which are adjacent to the link.

BGIS (BASIN GEOMORPHIC INFORMATION SYSTEMS)

The BGIS consist of interactive software for generating TIN-DEMs data structure and topographic analysis software which contain an automatic delineation of source areas to arbitrary part

Table 3.3. The Relation Between a Side-Attribute-Index and a Side-Component-Index

	Out-Flow-Side	Along-Flow-Side	In-Flow-Side
Out-Flow-Side	Valley segment	Valley segment	Slope segment
Along-Flow-Side	Valley segment	Slope segment	Ridge segment
In-Flow-Side	Slope segment	Ridge segment	Ridge segment

Table 3.4. Channel Network Data Set

No. of Link	No. of Downstream	No. of Upstream	No. of Vertices	Right Triangle	Left Triangle
I	Null	II, III	10, 11	c	b
			11, 12	m	f
II	I	Null	12, 13	l	k
			13, 14	r	s
III	I	Null	12, 15	j	g

of a channel network and mapping of a distribution of elevations, slopes, aspects, flow path lengths, and upslope contributing areas. A schematic outline of the BGIS is shown in Figure 3.4.

Source Data Sets

Source data sets are grid DEMs and DLGs of river courses. These data sets, produced by government agencies such as the United States Geological Survey (USGS) or the National Land Agency in Japan, are easily obtained. If source data sets for a particular study area are not available, they can be derived by digitizing contour lines and river courses on a topographic map by using a flatbed digitizer.

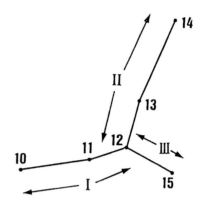

Figure 3.3. Sample channel network.

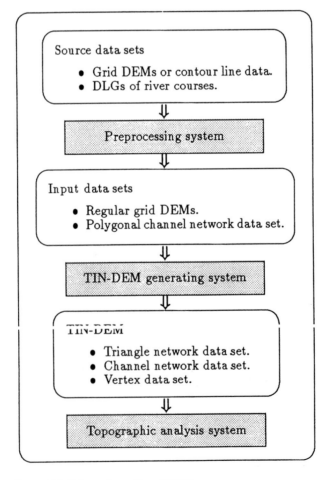

Figure 3.4. Schematic outline of BGIS.

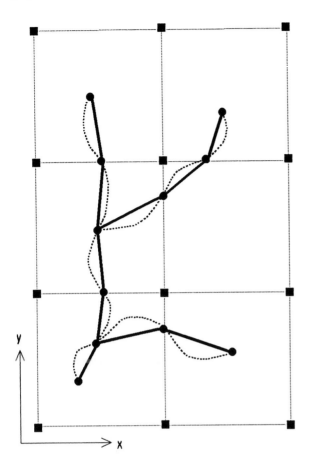

Figure 3.5. Schematic representation for making a polygonal channel network. Dashed lines denote a DLG of river courses. Solid lines denote a polygonal channel network.

Preprocessing System

From these source data sets, (1) a regular grid DEM, and (2) polygonal channel network data for a study area are produced. A regular grid DEM is interpolated from a grid DEM or contour line data. Polygonal channel network data are made up of polygonal lines which are derived by calculating the intersection of a straight line which connects two points on a regular grid DEM and a segment which connects two continuous points on a DLG of river courses (Figure 3.5).

TIN-DEMs Generating System

The two data sets made by the preprocessing systems are input into these systems and the three data sets noted in the TIN-DEMs DATA STRUCTURE section are generated. These systems include the following modules:

(a) a module for generating triangles from a regular grid DEM;
(b) a module for getting rid of pits;
(c) a module for joining discontinuous valley segments to a channel network; and
(d) a module for subdividing triangular facets.

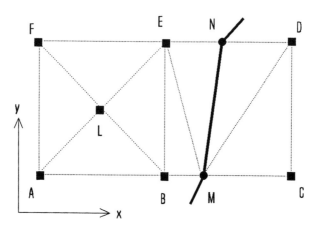

Figure 3.6. Automatic division of squares into triangles.

Module for Generating Triangles from a Regular Grid DEM

A data set which represents a basin with triangular facets based on a regular grid DEM and a polygonal channel data set are generated. For example, in Figure 3.6, the points A, . . . , F represent vertices on a grid DEM, and the segment MN represents a part of a channel network. For the square ABEF which has no channel segment in it, the point L is added in the center of the square, and it is subdivided to four triangles. The elevation of the added point L is interpolated using the elevations of four neighboring points. For the square BCDE which has one channel segment in it, it is subdivided to several triangles under the rule that the channel segment results in a side of a triangle. These two cases are processed automatically. In other cases, for example, a square which has more than one channel segment in it, and a square which has confluence points, upper ends of a channel network or a downstream end of a channel network in it, are subdivided using an interactive software. An operator can add new points if needed, make triangles manually, and see the result of a subdivision on a computer display. Figure 3.7 shows the example of a subdivision. The shaded area has already been subdivided into triangles. Squares which an operator needs to subdivide into triangles are not so many that the interactive handling of these subdivisions is not laborious and not time-consuming. After subdividing all the squares into triangles, side-attribute-indices, side-component-indices, and unit normal vectors of each triangular facet are computed, and the vertex data set, the triangle network data set, the channel network data set are produced.

Module for Getting Rid of Pits

A pit is a vertex whose surrounding vertices have higher elevations. If a natural topography is so complicated to represent it using a grid DEM with a current grid spacing, sometimes false pitting occurs. In this module, a pit is found automatically and solved by adding a new

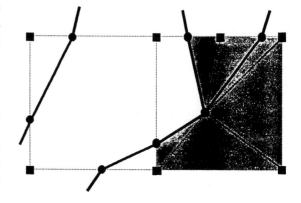

Figure 3.7. Interactive division of squares into triangles.

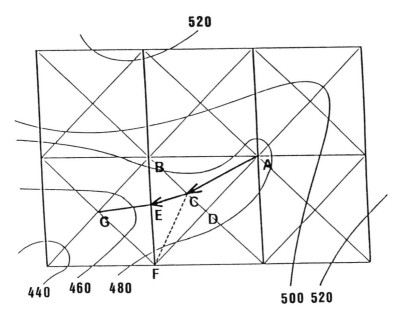

Figure 3.8. Schematic representation for getting rid of pits.

point and subdividing to triangles interactively. An algorithm for getting rid of a pit can be accomplished by following five steps:

Step 1: Find a vertex whose elevation is lower than the surrounding vertices (A, in Figure 3.8).
Step 2: Based on a topographic map, add a new point C, considering the direction of water
flow, and give an appropriate elevation to the point referring to a topographic map.
Step 3: Using the point C, subdivide triangle ABD to triangle ABC and triangle ACD, triangle
BFD to triangle BFC and triangle FDC.
Step 4: Update the vertex data set and the triangle network data set.
Step 5: If the new point C is a pit, return to step 1 and repeat
Step 1–5 until no false pits exist.

Module for Joining Discontinuous Valley Segments to a Channel Network

In a current model of a watershed basin, many valley segments exist. If these valley segments
do not join a channel network, the channel segment the triangles contribute to cannot be defined.
For example, for triangle bgk and triangle ikg in Figure 3.9, the channel segment they contribute
to cannot be defined. To correct this, the channel segment that the valley segments reach to is determined, after which, these valley segments are included in the channel network and the channel
network is reconstructed. An algorithm for this procedure is as follows:

Step 1: Find the lowest vertex in the continuous valley segments g in Figure 3.9.
Step 2: Trace the path of steepest descent from the lowest vertex until it reaches either a channel network or the boundary of the DEM.
Step 3: If it reaches to the channel networks, subdivide to triangles along the path of the steepest descent (in Figure 3.9, triangle ceg into triangle chg and triangle heg, triangle cde
into triangle cdh and triangle deh).
Step 4: Update the vertex data set and the triangle network data set.

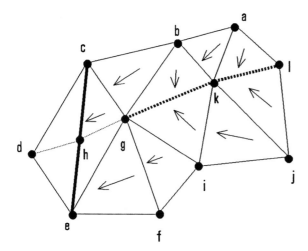

Figure 3.9. Schematic representation of discontinuous valley segment.

Step 5: Reconstruct a channel network and update the channel network data set. The channel networks before and after joining discontinuous valley to channel networks are shown in Figure 3.10.

Module for Subdivision of Triangles

Most of the triangles have two sides through which water flows out. To identify source areas, these triangles must be subdivided so that each triangle has only one out-flow-side contributing to only one adjacent triangle. An algorithm for this procedure is as follows:

Step 1: Trace a path of steepest ascent from a vertex, and find coordinates of an intersection on an opposite side.

Step 2: If the intersection is found on an opposite side (e on the segment bd in Figure 3.11), subdivide the triangle bcd to triangle bce and triangle cde, triangle abd to triangle abe and triangle aed.

Step 3: If the intersection exists on a ridge segment, stop. Otherwise continue until it encounters a ridge segment or a boundary of a TIN-DEM.

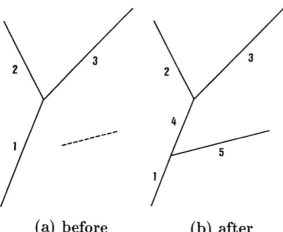

(a) before   **(b) after**

Figure 3.10. Reconstruction of channel network.

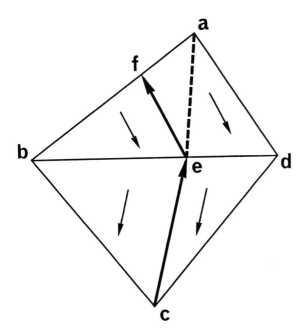

Figure 3.11. Subdivision to triangles which have one side through which water flows out.

This subdivision procedure is accomplished for all the vertices included in TIN-DEMs, but for new vertices added by this subdivision it is not necessary to apply this procedure.

APPLICATIONS

The BGIS was applied to three basins. Figure 3.12 shows a topographic map of the upper part of the Ara experimental basin. From this map, contour lines and a channel network were digitized manually by using a flatbed digitizer. Figure 3.13 shows the directions of water flow, ridges (bold solid lines), valleys (dashed lines), and the channel network (solid lines) for this study area. Figure 3.14 shows a three-dimensional representation of the basin, and the shaded areas represent the watershed basin delineated automatically.

Once the TIN-DEM data structure is generated, it is easy to identify source areas contributing to an arbitrary triangle. Each triangle has only one triangle which water flows into. When triangles which contribute to a particular triangle are found, a triangle from which water flows into it is found and added to a list of source areas recursively, until a triangle which has two ridge sides, or one ridge side and one along side, is added to the list. The channel network data set includes the numbers of the triangles which contact with a channel network, so by beginning this procedure with these triangles, all the triangles included in the watershed are identified.

Figures 3.15 and 3.16 show the Ara experimental basin (0.184 km^2) and the Ina basin (54.0 km^2). The number of vertices and triangles after processed by each module are represented in Table 3.5.

CONCLUSIONS

Geographic information systems in hydrologic modeling, the BGIS (Basin Geomorphic Information Systems) were presented for modeling a river basin using a TIN-DEM data structure. The BGIS are made up of interactive software for generating three data sets, (1) a vertex data set, (2) a triangle network data set, and (3) a channel network data set, and includes topographic analysis

Figure 3.12. Topographic map for the upper part of the Ara experimental basin.

Figure 3.13. Directions of water flow of the upper part of the Ara experimental basin.

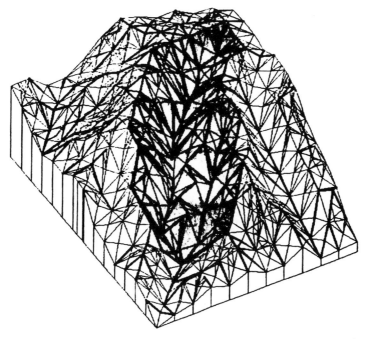

Figure 3.14. TIN-DEM for the upper part of the Ara experimental basin.

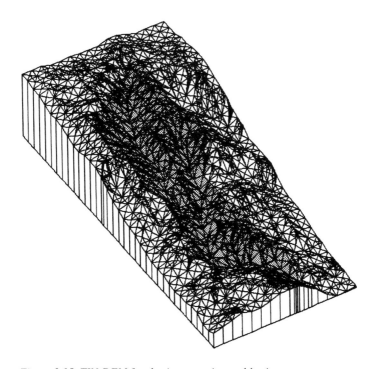

Figure 3.15. TIN-DEM for the Ara experimental basin.

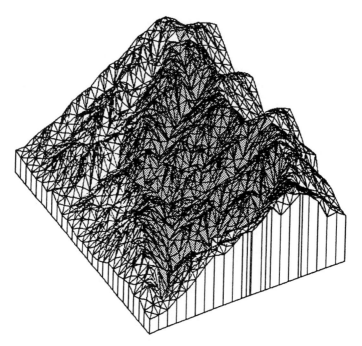

Figure 3.16. TIN-DEM for the Ina basin.

Table 3.5. The Number of Triangles and Vertices after Processed by Each Module

Module	Ina Basin		Ara Experimental Basin	
	No. of Vertices	No. of Triangles	No. of Vertices	No. of Triangles
Module for generating triangles from a regular grid DEM	1123	2146	1058	2016
Module for getting rid of pits	1151	2202	1071	2042
Module for joining discontinuous valley segments to a channel network	1185	2240	1100	2100
Module for subdividing triangular facets	3900	7682	3304	6483
Grid spacing	250 m		25 m	
Number of pits	5		2	

software which makes an automatical delineation of source areas to an arbitrary part of a channel network and mapping of a distribution of elevations, slopes, aspects, flow path lengths, and up-slope contributing areas.

This TIN-based topographic model incorporates the advantages of grid-based methods with their ease of computational implementation, efficiency, and availability of topographic databases, and combines the advantages of contour-based methods such as subdivision of a basin considering the direction of water flow. This form of discretization is advantageous for modeling water movement of a natural watershed basin.

REFERENCES

Band, L., 1986. Topographic partition of watersheds with Digital Elevation Models. *Water Resources Research*, 22(1):15–24.

Hutchinson, M. F., 1989. A new procedure for gridding elevation and stream line data with automatic removal of spurious pits. *Journal of Hydrology*, 106:211–232.

Jett, S. C., A. A. Weeks, W. M. Grayman, and W. E. Gates, 1979. Geographic Information Systems in hydrologic modeling. In: *Hydrologic Transport Modeling Symposium*, 10–11 December 1979, A.S.A.E, New Orleans, LA, pp. 127–137.

Jones, N. L., S. G. Wright, and D. R. Maidment, 1990. Watershed delineation with triangle-based terrain models, *J. Hydraul. Eng.*, 116(10):1232–1251.

Lu, M., T. Koike, and N. Hayakawa, 1989. A rainfall-runoff model using distributed data of radar rain and altitude. *Proceedings of JSCE*, 411(2–12):135–142.

Moore, I. D., E. M. O'Loughlin, and G. J. Burch, 1988. A contour-based topographic model for hydrological and ecological application. *Earth Surface Processes and Landforms*, 13:305–320.

Moore, I. D., and G. R. Foster, 1990. Hydraulics and overland flow. In *Process Studies in Hillslope Hydrology*, Anderson, M. G. and T. P. Burt, Eds., John Wiley and Sons, pp. 215–254.

Moore, I. D., R. B. Grayson, and A. R. Ladson, 1991. Digital terrain modelling : A review of hydrological, geomorphological and biological applications. *Hydrological Process*, 5(1):3–30.

O'Callaghan, J. F., and D. M. Mark, 1984. The extraction of drainage networks from digital elevation data. *Computer Vision Graphics and Image Processing*, 28:323–344.

O'Loughlin, E. M., 1986. Prediction of surface saturation zones in natural catchments by topographic analysis. *Water Resources Research*, 22(5):794–804.

Palacios-Velez, O. L., and B. Cuevas-Renaud, 1986. Automated river-course, ridge and basin delineation from digital elevation data. *Journal of Hydrology*, 86:299–314.

Palacios-Velez, O. L., and B. Cuevas-Renaud, 1991. SHIFT : A distributed runoff model using irregular triangular facets. *Journal of Hydrology*, 134:35–55.

Quinn, P., K. Beven, P. Chevallier, and O. Planchon, 1991. The prediction of hillslope flow paths for distributed hydrological modelling using digital terrain models. *Hydrological Process*, 5:59–79.

Takara, K., and T. Takasao, 1991. Fractal dimension of river basins based on digital terrain maps. *Proceedings of Hydraulic Engineering*, Japan Society of Civil Engineers, 35:135–142.

Takasao, T., and K. Takara, 1989. A basic analysis of geomorphologic features of drainage basins and channel networks based on the digital national land information. *Annals of the Disaster Prevention Research Institute*, Kyoto University, 32(B–2), pp. 435–454.

Tarboton, D. G., R. L. Bras, and I. Rodriquez-Iturbe, 1989. The Analysis of River Basins and Channel Networks Using Digital Terrain Data. Dept. of Civil Engineering, M.I.T, TR No. 326, Cambridge, MA.

Vieux, B. E., 1991. Geographic Information Systems and non-point source water quality and quantity modeling. *Hydrological Process*, 5:101–113.

Wyss, J., E. R. Williams, and R. L. Bras, 1990. Hydrologic modeling of England river basins using radar rainfall data. *Journal of Geophysical Research*, 95(D3):2143–2152.

CHAPTER 4

Basinsoft, a Computer Program to Quantify Drainage Basin Characteristics

Craig A. Harvey and David A. Eash

INTRODUCTION

Surface water runoff is a function of many interrelated factors including climate, soils, land-use, and the physiography of the drainage basin. A practical and effective method to quantify drainage basin characteristics would allow analysis of the interrelations of these factors, leading to an improved understanding of the effects of drainage basin characteristics on surface-water runoff. Historically, the quantification of drainage basin characteristics has been a tedious and time-consuming process. Recent improvements in computer hardware and software technology have enabled the developers of a program called Basinsoft to automate this process. Basinsoft requires minimal preprocessing of data and provides an efficient, automated procedure for quantifying selected morphometric characteristics and the option to area-weight characteristics for a drainage basin. The user of Basinsoft is assumed to have a limited amount of experience in the use of ARC/INFO, a proprietary geographic information system (GIS). (The use of brand names in this chapter is for identification purposes only and does not constitute endorsement by the U.S. Geological Survey [USGS].)

In 1988, the USGS began developing a program called Basinsoft. The initial program quantified 16 selected drainage basin characteristics from three source-data layers that were manually digitized from topographic maps using the versions of ARC/INFO, Fortran programs, and prime system Command Programming Language (CPL) programs available in 1988 (Majure and Soenksen, 1991). By 1991, Basinsoft was enhanced to quantify 27 selected drainage-basin characteristics from three source-data layers automatically generated from digital elevation model (DEM) data using a set of Fortran programs (Majure and Eash, 1991; Jenson and Dominique, 1988). Due to edge-matching problems encountered in 1991 with the preprocessing of the DEM data, the Basinsoft program was subsequently modified to quantify 24 selected drainage-basin characteristics from four source-data layers created from three types of data (topographic maps, digital line graph [DLG] data, and DEM data) (Eash, 1993, 1994). The early versions of Basinsoft relied primarily on Fortran programs and prime system CPL to manage data and calculate statistics, thus making them platform dependent.

In 1994, Basinsoft was redeveloped entirely using Arc Macro Language (AML), a postprocessing language written to run in ARC/INFO, to increase portability among systems using ARC/INFO version 7.0.2 or later (Environmental Systems Research Institute, 1994). The current (1997) version of Basinsoft processes four source-data layers representing selected aspects of a drainage basin to quantify selected morphometric drainage basin characteristics (Harvey and Eash, 1996). The 27 selected basin characteristics quantified by Basinsoft are listed in Table 4.1 as meas-

Table 4.1. Selected Drainage Basin Characteristics Quantified Using Basinsoft

Basin-Area Quantifications

TDA—Total drainage area[a,b], in square miles, an internal measurement maintained by ARC/INFO. TDA is acquired from the drainage-divide data layer (cover_bas) attribute table as the area measurement and it includes noncontributing areas.

NCDA—Noncontributing drainage area[a], in square miles, is the total area that does not contribute to surface-water runoff at the basin outlet. NCDA is obtained by computing summary statistics on the drainage-divide data layer (cover_bas) attribute table based on the item CONTRIB.

Basin-Length Quantifications

BL—Basin length[a], in miles, measured along a line areally centered through the drainage-divide data layer (cover_bas) from basin outlet to where the main channel extended meets the basin divide. This process uses ARC/GRID to calculate the centerline.

BP—Basin perimeter[a], in miles, measured along entire drainage-basin divide. BP is an internal measurement maintained by ARC/INFO and is acquired from the drainage-divide data layer (cover_bas).

Basin-Relief Quantifications

BS—Average basin slope[a,b], in feet per mile, measured by the "contour-band" method, within the contributing drainage area (CDA). Summary statistics are performed on the hypsography data layer (cover_con). The output from the statistics command is used in conjunction with the user-designated elevation-contour interval as input into the formula to calculate BS. BS = (total length of all selected elevation contours) (contour interval) / CDA.

BR—Basin relief[c], in feet, measured as the difference between the elevation of the highest grid cell and the elevation of the grid cell at the basin outlet. BR uses the lattice (grid) data layer (cover_lat) to determine the minimum elevation as the land-surface elevation at the basin outlet. The maximum elevation is determined from the lattice data layer (cover_lat) statistics INFO file item ZMAX.

Basin-Aspect Quantifications

BA—Basin azimuth[a,b], in degrees, compass direction of a line projected from where the main channel, if extended, meets the basin divide downslope to the basin outlet. Measured clockwise from north at 0°.

Basin Computations

CDA—Contributing drainage area[a,b], in square miles, defined as the total area that contributes to surface-water runoff at the basin outlet, CDA = TDA − NCDA.

BW—Effective basin width[a], in miles, BW = CDA / BL.

SF—Shape factor[a], dimensionless, ratio of basin length to effective basin width, SF = BL / BW.

ER—Elongation ratio[a], dimensionless, ratio of (1) the diameter of a circle of area equal to that of the basin to (2) the length of the basin, ER = $[4\,CDA / \pi\,(BL)^2]^{0.5}$ = 1.13 $(\frac{1}{SF})^{0.5}$.

RB—Rotundity of basin[a], dimensionless, RB = $[\pi\,(BL)^2] / [4\,CDA]$ = 0.785 SF.

CR—Compactness ratio[a], dimensionless, the ratio of the perimeter of the basin to the circumference of a circle of equal area, CR = $BP/2\,(\pi\,CDA)^{0.5}$.

RR—Relative relief[d], in feet per mile, RR = BR / BP.

Channel- or Stream-Length Quantifications

MCL—Main channel length[a,b], in miles, measured along the main channel from the basin outlet to where the main channel, if extended, meets the basin divide. Summary statistics are computed on the hydrography data layer (cover_str) based on the item CODE.

TSL—Total stream length[c], in miles, computed by summing the length of all stream segments within the CDA using summary statistics on the hydrography data layer (cover_str) based on the item LENGTH.

Table 4.1. Selected Drainage-Basin Characteristics Quantified Using Basinsoft (Continued)

Channel-Relief Measurement

MCS—Main-channel slope[a,b], in feet per mile, an index of the slope of the main channel computed from the difference in streambed elevation at points 10% and 85% of the distances along the main channel from the basin outlet to the basin divide. A route system is developed based on the INFO-item CODE equal to 1 in the hydrography (cover_str) data layer. The 10% and 85% distances from the basin outlet are calculated and nodes are placed at those positions along the route. The nodes are converted to points and elevations are determined for each point from the lattice data layer (cover_lat) and attributed to a temporary data layer for use in the MCS formula. $MCS = (E_{85} - E_{10}) / (0.75\ MCL)$.

Channel or Stream Computations

MCSR—Main-channel sinuosity ratio[a], dimensionless, $MCSR = MCL / BL$.

SD—Stream density[a], in miles per square mile, within the CDA, $SD = TSL / CDA$.

CCM—Constant of channel maintenance[a], in square miles per mile, within the CDA, $CCM = CDA / TSL = 1 / SD$.

MCSP—Main channel slope proportion[c], dimensionless, $MCSP = MCL / (MCS)^{0.5}$.

RN—Ruggedness number[e], in feet per mile, $RN = (TSL)\ (BR) / CDA = (SD)\ (BR)$.

SR—Slope ratio of main-channel slope to basin slope[c], dimensionless, within the CDA, $SR = MCS / BS$.

Stream-Order Quantifications

FOS—Number of first-order streams within the CDA[f], dimensionless. FOS is computed using Strahler's method of stream ordering and summary statistics on the hydrography data layer (cover_str).

BSO—Basin Stream Order[f], dimensionless, stream order of the main channel at the basin outlet. BSO is computed by intersecting the main channel with the drainage-divide data layer and determining the Strahler-stream order of the stream at the basin outlet.

Stream-Order Computations

DF—Drainage frequency[c], in number of first-order streams per square mile, within the CDA, $DF = FOS / CDA$.

RSD—Relative stream density[g], dimensionless, within the CDA, $RSD = (FOS)(CDA)/(TSL)^2 = DF/(SD)^2$.

Footnotes reference the literary sources for drainage-basin charactertistics:
[a]Modified from Office of Water Data Coordination (1978, pp. 7–9—7–16).
[b]Modified from National Water Data Storage and Retrieval System (Dempster, 1983, pp. A–24–A–26).
[c]Modified from Strahler (1958, pp. 282–283 and 289).
[d]Modified from Melton (1957).
[e]Modified from Robbins (1986, p. 12).
[f]Modified from Strahler (1952, p.1120).
[g]Modified from Melton (1958).

urements or computations. Twelve of the drainage-basin characteristics constitute specific quantifications of area, length, relief, aspect, and stream order; the other 15 characteristics are computations that make use of the various quantifications to calculate other drainage basin characteristics or statistics.

Development of the Basinsoft program is continuing. Information presented in this chapter on Basinsoft is republished in a slightly condensed and modified version (reflects revisions to the current version of Basinsoft as of July 1997) from the report "Description, Instructions, and Verification for Basinsoft, a Computer Program to Quantify Drainage-Basin Characteristics" (Harvey and Eash, 1996). Information regarding the status of this program may be obtained by accessing the PixSell Home Page at http://www.pixsell.com or by sending an email to basinsoft@pixsell.com or by contacting PixSell, Bldg. 2105, John C. Stennis Space Center, Bay St. Louis, MS 39529.

Table 4.2. "Core" AMLs and Associated Functions

AML Name	AML Function[a]
basinsoft.aml	Driver AML, initiates other core AMLs.
basinit.aml	Initializes variables and creates an empty INFO table to hold drainage basin characteristics as they are quantified by Basinsoft.
bam.aml	Quantifies TDA, NCDA, and CDA.
clm.aml	Quantifies MCL and TSL.
azim.aml	Quantifies BP and BA.
orient.aml	Calculates and sets the global variable ".orient".
basinl.aml	Quantifies BL.
dfm.aml	Quantifies FOS, BSO, and DF.
brm.aml	Quantifies BS and BR.
crm.aml	Quantifies MCS.
computations.aml	Quantifies BW, SF, ER, RB, CR, RR, MCSR, SD, CCM, MCSP, RN, SR, and RSD.

[a]See Table 4.1 for description of drainage-basin characteristic quantifications.

BASINSOFT PROGRAM STRUCTURE

The AML programs comprising Basinsoft are initialized so that each program performs specific tasks and when necessary, passes information required by the following program for subsequent processing. By programming in this modular manner, additional or future modules are easily appended to Basinsoft by the user.

The "core" AML programs comprising Basinsoft, listed with their associated functions in Table 4.2, are referred to as modules in this chapter. Basinsoft.aml controls the modules found in the Basinsoft directory (Figure 4.1) and can be modified to control some of the utility programs found in the tools directory (Figure 4.1). Modules are executed each time Basinsoft is initiated. Drainage-basin characteristics are quantified within the modules. Examples of the directory structure required by Basinsoft are listed in Table 4.3 and shown in Figure 4.2.

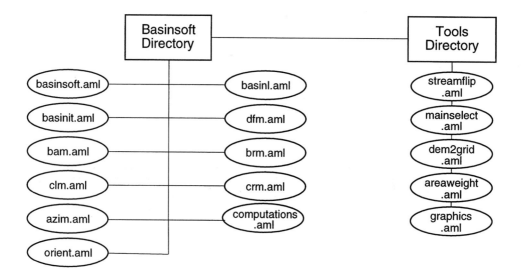

Figure 4.1. Example of Basinsoft program directory structure.

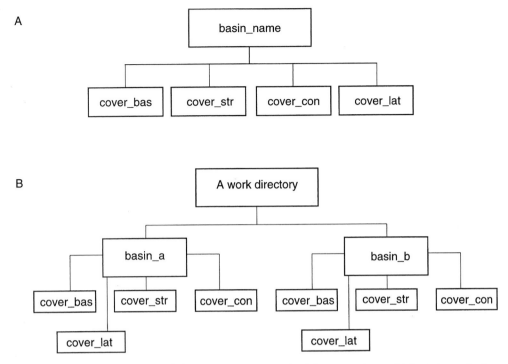

Figure 4.2. Naming conventions and directory structure required for processing multiple drainage basins—(A) generic directory structure of single basin, (B) actual directory structure for multiple basins (e.g., basin_a and basin_b) maintained under a work directory.

COMPUTER RESOURCES REQUIRED TO EXECUTE BASINSOFT

Basinsoft was developed and tested on Data General Avion model 300, 530, and 550 workstations using ARC/INFO version 7.0.2, and was subsequently ported to run on NT with ARC/INFO version 8.1. Basinsoft requires no compilation. Processing time to execute Basinsoft varies between 5 and 20 minutes on a Data General Avion 530 for drainage basins with areas less than 3,000 square miles. Processing time on other systems may vary depending on their hardware capabilities. These processing times do not include time for preprocessing, which will vary according to the type of source data available for generating the four source-data layers required as input to Basinsoft. The time to execute the TOPOGRID command may range from 5 minutes to more

Table 4.3. Naming Conventions and Description of Directory Structure Required by Basinsoft

Directory Name	Description of Directory
basin_name	This is the unique name defining a specific drainage basin (e.g., basin_a). The following data layers are source-data layers maintained under the basin_name directory.
cover_bas	Drainage-divide data layer.
cover_str	Hydrography data layer.
cover_con	Hypsography data layer.
cover_lat	Lattice data layer with areal extent clipped to drainage-basin boundary.

than 1 hour, depending on the size of the drainage basin and hardware capabilities. The basin-length module also is computer-time, memory, and disk-space intensive.

SOURCE-DATA REQUIREMENTS OF BASINSOFT

Basinsoft requires the user to provide four source-data layers. Figure 4.3 (see color section) shows examples of the four source-data layers, which are: (1) a drainage-divide (Figure 4.3A), (2) hydrography (Figure 4.3B), (3) hypsography (Figure 4.3C), and (4) a lattice elevation model (Figure 4.3D). These data layers are not required to be of any specific scale; however, the hydrography and hypsography data layers should be clipped to the areal extent of the drainage-divide data layer. The lattice elevation model is preprocessed into a lattice data layer—cover_lat, with the same areal extent as the drainage-divide data layer.

Several programs were developed to assist the user in preprocessing source-data layers in preparation for use with Basinsoft. These programs, which are found in the tools directory, perform various tasks such as creating a hydrologically accurate lattice elevation model, facilitating the creation of a hypsography data layer, and converting a 1:250,000-scale DEM to a lattice data layer. Several programs were developed to assist the user in assigning attributes to data layers according to a specialized attribute scheme required by Basinsoft.

PREPROCESSING OF SOURCE DATA FOR PROGRAM INPUT

Basinsoft was developed to run with minimal preprocessing; however, there are several steps required to prepare the source-data layers for subsequent processing by Basinsoft. Preprocessing steps include the generation of four source-data layers as described above and the assignment of attributes. The arc segments of the hydrography data layer must be edited using the streamflip.aml to orient the arc segments in a downstream direction. The main channel of the drainage basin must be delineated, attributed, and extended to the drainage-divide. AMLs were developed to assist the user with nearly all aspects of the preprocessing required by Basinsoft.

Preprocessing the Drainage-Divide Data Layer

The drainage-divide data layer (cover_bas) (Figure 4.3A) must be attributed with an INFO-item named CONTRIB. This item should have a value of '0' (the default) for contributing drainage areas (CDA) and a value of '1' for noncontributing drainage areas (NCDA); thus no attribute generation is necessary unless NCDA exist. A second INFO-item, BASINNAME, must be added to the drainage-divide data layer and attributed with an identifying name for the drainage basin.

Preprocessing the Hydrography Data Layer

The topology of the surface-water drainage network or hydrography data layer (cover_str) (Figure 4.3B) must be clean, and the main channel must be delineated and attributed with an INFO-item named CODE. The arc segments in the hydrography data layer (cover_str) must be oriented in a downstream direction. AML programs named streamflip.aml and mainselect.aml were developed to assist in performing these tasks and specialized attributing.

Streamflip.aml

The number of first-order streams (FOS) within a drainage basin is used to quantify three drainage basin characteristics. The number of first-order streams is determined within Basinsoft using ARC/INFO's STRAHLER program. The STRAHLER program requires that all FROM/TO nodes of arc segments representing the hydrography be oriented in a downstream direction. The streamflip.aml was developed to help the user accomplish this editing. The AML will orient the FROM/TO nodes of each arc segment in a downstream direction.

Mainselect.aml

The main channel length (MCL) is used to quantify four drainage basin characteristics. By definition, MCL is measured along the main channel from the basin outlet to where the main channel, if extrapolated, were to meet the basin divide. The quantification of MCL requires the delineation or attribution of the main channel and, in most instances, the addition of an arc segment from the upstream node of the main channel to the drainage basin divide. An automated procedure named mainselect.aml was developed to assist in extending and defining the main channel. The mainselect.aml requires arc segments to be oriented in a downstream direction. The AML will assign a value of '1' to the INFO-item named CODE for arcs which represent part of the main channel, and a value of '0' (the default) to all other arcs in the hydrography data layer (cov_str).

Preprocessing the Hypsography Data Layer

Elevation contour or hypsography data may be available in the form of DLG data or may need to be generated from DEM data. Other sources of data, such as triangular irregular network (TIN) data or digitized vector data, may be used to generate elevation contour data.

Several steps are necessary if the hypsography data layer (cover_con) (Figure 4.3C) is to be created from DEM data. A utility AML, dem2grid.aml, located in the tools directory, was developed to convert DEM (raster) data to hypsography (vector) data. This task also can be accomplished interactively using standard ARC/INFO commands.

Preprocessing the Lattice Data Layer

If the hypsography data layer is developed using DEM data, and the utility AML dem2grid.aml is used to generate hypsography data, then the lattice elevation model (Figure 4.3D) is a by-product of executing dem2grid.aml. A lattice data layer needs to be generated. The lattice data layer, cover_lat, will have its areal extent defined by the drainage-divide data layer (cover_bas). An optional module, dem2grid.aml was developed to create the lattice data layer and the hypsography data layer (cover_con) from DEM data.

The elevation source data (lattice elevation model) (Figure 4.3D) may be unavailable, or may be available at an inappropriate scale. Under these circumstances it may be necessary to create or enhance the elevation data. One method is to use the ARC/INFO command TOPOGRID to create the lattice data layer. This command uses hypsography and a single-line hydrography network with options to use point elevation data and linear data as break lines in the topography to create a hydrologically accurate lattice data layer.

PROCESSING DATA LAYERS TO QUANTIFY DRAINAGE BASIN CHARACTERISTICS

Basinsoft processing can be performed on either single or multiple basins. The process of quantifying basin characteristics for a number of drainage basins is highly repetitive; therefore, it is possible to automate much of this task once the preprocessing has been completed. Processing involves invoking the program Basinsoft.aml to quantify 27 selected morphometric drainage basin characteristics.

A single drainage basin can be processed from the ARC prompt. Basinsoft is executed with the appropriate basin name as a command line argument. Upon execution of Basinsoft, the drainage basin is processed and selected output is written into the basin_name directory (Figure 4.2A).

Multiple drainage-basin processing can be performed using the program multi_p.aml, an AML provided in the tools directory. The user modifies the AML to include a list of names of drainage basins to be processed. Strict adherence to the naming convention defined in Table 4.3 is necessary to automate this procedure. Upon execution, the program multi_p.aml loops through the list of basin names provided by the user, and output files are written into the basin_name directory (Figure 4.2B). The directory structure, as shown in Figure 4.2B, is needed to execute the program multi_p.aml.

BASINSOFT OUTPUT

Basinsoft output is generated in two formats. An INFO table named COVER_CHAR and a flat file named CHARS.TXT are generated in the basin_name directory (Table 4.3). The flat file is generated from the INFO table. Both files are output with the following format: characteristic_name = value. Table 4.4 shows an example output in the flat-file format. An optional graphic-output module named graphics.aml is available in the tools directory.

Table 4.4. Example of Flat File (ASCII) Output from Basinsoft

Basin Name			=		beavtr
TDA	=	11.890	MCL	=	5.689
NCDA	=	0.000	TSL	=	15.086
CDA	=	11.890	MCS	=	23.381
BL	=	5.271	MCSR	=	1.079
BP	=	15.528	SD	=	1.269
BS	=	53.549	CCM	=	0.788
BR	=	151.246	MCSP	=	1.176
BA	=	55.063	RN	=	191.896
BW	=	2.256	SR	=	0.437
SF	=	2.336	FOS	=	6.000
ER	=	0.739	BSO	=	3.0
RB	=	1.834	DF	=	0.505
CR	=	1.270	RSD	=	0.313
RR	=	9.740			

BASINSOFT VERIFICATION

To verify the accuracy of the drainage basin characteristics quantified using Basinsoft, manual measurements of 12 selected drainage basin characteristics were made from USGS topographic maps for drainage areas upstream of 11 streamflow-gauging stations in Iowa. Manual measurements were made at scales identical to the quantifications done by Basinsoft. The results of the comparisons are shown in Table 4.5.

The Wilcoxon signed-ranks test was applied to determine the statistical significance between median manual measurements and median Basinsoft quantifications of 10 of the 12 drainage basin characteristics listed in Table 4.5. No test was performed for noncontributing drainage area

Table 4.5. Comparisons of Manual Measurements and Basinsoft Quantifications of Selected Drainage Basin Characteristics for 11 Streamflow Gauging stations in Iowa[a]

Station Number	Measurement Technique	Drainage Basin Characteristic											
		TDA[b]	NCDA	BL	BP	BS	BR	BA	MCL	TSL	MCS	FOS	BSO
05411600	MAN	177	0	27.0	73.3	166	297	144	36.4	242	5.61	84	4
	BSOFT	178	0	25.6	73.9	73.6	280	143	36.3	236	6.14	84	4
	% DIFF	+0.6	0	−5.2	+0.8	−55.7	−5.7	−0.7	−0.3	−2.5	+9.4	0	0
05414450	MAN	21.6	0	8.81	21.9	426	444	104	11.4	31.9	19.1	10	3
	BSOFT	22.3	0	8.47	21.3	264	439	104	11.1	31.5	23.1	10	3
	% DIFF	+3.2	0	−3.9	−2.7	−38.0	−1.1	0	−2.6	−1.3	+20.9	0	0
05414600	MAN	1.54	0	2.31	5.32	246	280	68	2.63	2.63	101	1	1
	BSOFT	1.53	0	2.10	5.97	208	299	75	2.58	2.58	108	1	1
	% DIFF	−0.6	0	−9.1	+12.2	−15.4	+6.8	+10.3	−1.9	−1.9	+6.9	0	0
05462750	MAN	11.6	0	4.84	15.0	157	160	52	5.74	15.2	28.3	6	3
	BSOFT	11.9	0	5.27	15.5	53.5	151	55	5.69	15.1	23.4	6	3
	% DIFF	+2.6	0	+8.9	+3.3	−65.9	−5.6	+5.8	−0.9	−0.7	−17.3	0	0
05463090	MAN	56.9	0	11.6	33.5	ND	181	91	17.4	73.5	7.27	28	3
	BSOFT	57.0	0	11.4	33.1	52.3	187	89	16.8	73.9	8.67	28	3
	% DIFF	+0.2	0	−1.7	−1.2	ND	+3.3	−2.2	−3.4	+0.5	+19.3	0	0
05470500	MAN	204	0	24.4	69.8	99.6	318	150	37.8	210	7.52	60	4
	BSOFT	208	0	25.3	67.7	49.0	309	151	35.7	192	7.81	51	4
	% DIFF	+2.0	0	+3.7	−3.0	−50.8	−2.8	+0.7	−5.6	−8.6	+3.9	−15.0	0
05481000	MAN	844	0	51.5	139	ND	303	175	88.1	685	2.04	152	5
	BSOFT	852	0	52.3	139	33.4	269	176	88.9	685	1.50	155	5
	% DIFF	+0.9	0	+1.6	0	ND	−11.2	+0.6	+0.9	0	−26.5	+2.0	0
05489490	MAN	22.9	0	10.5	24.8	289	280	70	13.3	28.4	14.8	10	2
	BSOFT	22.2	0	10.6	26.2	165	286	68	13.0	27.6	20.2	10	2
	% DIFF	−3.1	0	+1.0	+5.6	−42.9	+2.1	−2.9	−2.3	−2.8	+36.5	0	0
06609500	MAN	871	0	80.9	206	352	537	201	101	1,230	3.18	477	5
	BSOFT	869	0	81.7	210	197	491	203	99.9	1,270	3.34	487	5
	% DIFF	−0.2	0	+1.0	+1.9	−44.0	−8.6	+1.0	−1.1	+3.3	+5.0	+2.1	0
06807780	MAN	42.7	0	21.1	47.4	346	268	204	22.2	52.7	9.37	18	3
	BSOFT	42.8	0	21.5	48.8	195	283	205	22.2	55.3	10.1	19	3
	% DIFF	+0.2	0	+1.9	+3.0	−43.6	+5.6	+0.5	0	+4.9	+7.8	+5.6	0
06903400	MAN	182	0	21.9	79.0	152	224	57	39.5	228	3.24	80	4
	BSOFT	184	0	21.3	79.6	82.2	200	57	39.6	231	3.37	80	4
	% DIFF	+1.1	0	−2.7	+0.8	−45.9	−10.7	0	+0.3	+1.3	+4.0	0	0
WSRT p-value[c]		0.1192	NT	0.6248	0.2125	0.0092	0.2296	0.3447	0.0908	0.9291	0.1000	0.3742	NT

[a]TDA, total drainage area, in square miles; NCDA, noncontributing drainage area, in square miles; BL, basin length, in miles; BP, basin perimeter, in miles; BS, average basin slope, in feet per mile; BR, basin relief, in feet; BA, basin azimuth, in degrees; MCL, main channel length, in miles; TSL, total stream length, in miles; MCS, main channel slope, in feet per mile; FOS, number of first-order streams; BSO, basin stream order; MAN, manual measurement; BSOFT, Basinsoft quantification; % DIFF, percentage difference between MAN and BSOFT; ND, not determined; WSRT, Wilcoxon signed-ranks test; NT, no test performed because all values for % DIFF = 0.

[b]Manual TDA measurements are streamflow gauging station drainage areas published by the U.S. Geological Survey in annual streamflow reports.

[c]In general, p-values greater than 0.05 indicate that there is not a statistically significant difference between the median of the manual measurements and the median of the Basinsoft quantifications, using a 95% level of significance for a two-tailed Wilcoxon signed-ranks test.

(NCDA) and basin stream order (BSO) because these characteristics either were equal to zero or there was no difference between manual measurements and Basinsoft quantifications for all 11 drainage basins. Results of the statistical comparison tests indicate that Basinsoft quantifications were not significantly different (p-value >0.05) from manual measurements for 9 of the 10 drainage basin characteristics tested (Table 4.5). Basin slope (BS) was the only characteristic tested for which Basinsoft quantifications were significantly different (p-value <0.05) from manual measurements.

The results of a comparison test for average basin slope (BS) using three methods of measuring elevation contour lengths are listed in Table 4.6. The results indicated that Basinsoft quantifies basin slope with acceptable results; however, ARC/INFO was unable to generate appropriate elevation contours from 1:250,000-scale DEM data for comparison with manual measurements of elevation contours from 1:250,000-scale topographic maps. Comparisons for basin slope (Tables 4.5 and 4.6) appeared to indicate that the 1:250,000-scale DEM data were too coarse for ARC/INFO to accurately reproduce elevation contour data with the sinuosity found on the 1:250,000-scale topographic maps. Figure 4.4 shows elevation contours generated from DEM data using ARC/INFO (Figure 4.4A) are much more generalized than elevation contours digitized from topographic maps of the same scale (Figure 4.4B). Thus, the total length of contours generated from DEM data are underrepresented when compared to contours shown on topographic maps (Table 4.6). This contour overgeneralization illustrates how the total length of the elevation contours are underestimated by Basinsoft using the "contour-band" method to quantify basin slope (Table 4.1).

Understanding the reason for differences between the manual measurements and quantifications made by Basinsoft is important in determining the type of comparisons which may be relevant in analysis of this type of data. The data in Table 4.6 indicated that it was not preferable to compare Basinsoft quantifications of contour data generated from DEM data by ARC/INFO with either digitized contours or contours generated from DLG data; however, measurements based on elevation contours digitized from contour maps and those based on elevation contours generated from DLG data were similar.

Basinsoft quantifications of main channel slope (MCS) have the greatest range in percent dif-

Table 4.6. Comparison of Elevation-Contour Length Measurements Used to Quantify Basin Slope

Source Data (All Data Were 1:250,000-scale)	Method of Measurement	Elevation-Contour Length (CL) Mesurement, in Miles	Contributing Drainage Area (CDA), in Square Miles	Contour Interval (CI), in Feet	Average Basin Slope Quantified Using CL * CI/CDA, in Feet per Mile
Topographic map	Manual measurement of elevation contours from topographic map	37.23	11.89	50	156.6
Topographic map (Figure 4.4B)	Basinsoft quantification of elevation contours digitized from topographic map	35.05	11.89	50	147.4
DEM data (Figure 4.4A)	Basinsoft quantification of elevation contours generated using ARC/INFO	12.73	11.89	50	53.5

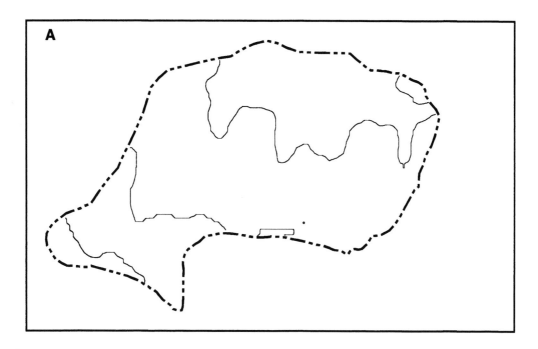

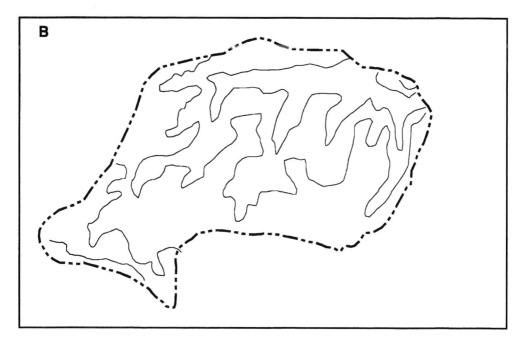

Figure 4.4 A-B. Elevation contours generated (A), with a 50-foot interval from 1:250,000–scale DEM data using ARC/INFO and (B), with a 50–foot interval digitized from a 1:250,000–scale topographic map.

ference between manual measurements and Basinsoft quantifications of the 10 selected drainage basin characteristics listed in Table 4.5. There are two main components of the MCS equation, length and elevation (Table 4.1). Quantification differences for MCS ranged from –26.5% to +36.5% (Table 4.5). However, quantification differences for main channel length (MCL) ranged from –5.6% to +0.9% (Table 4.5). These quantification differences indicated that the variation between manual measurements and Basinsoft quantifications of MCS were due mainly to differences in the determination of the streambed elevation at points 10% and 85% of the distance along the main channel from the basin outlet.

The approximate time required for an experienced ARC/INFO user to process three of the drainage basins listed in Table 4.5 using Basinsoft was 8 hours. The three basins represent large, intermediate, and small drainage areas. Manual measurements made from the same scales of topographic maps as used by Basinsoft for these three drainage basins required approximately 50 hours.

Preliminary comparisons between basin characteristic measurements made from various scales of cartographic data for these three drainage basins appear to indicate that several of the basin characteristics (BS, MCL, TSL, MCS, FOS, BSO) are map-scale dependent (Eash, 1993; Eash, 1994). Map-scale dependency refers to the effect on a measurement when that measurement is made from a cartographic data source of some specific scale as compared to that same measurement made from a different scale. Thus, interbasin comparisons of Basinsoft quantifications may be unreliable if different scales of cartographic data, or different sources of digital data (that is, raster versus vector), are used in generating the four source-data layers for each drainage basin.

Source-data layers obtained from larger-scale cartographic data and processed using Basinsoft may provide the best drainage basin quantifications for a study area. However, the scales and sources of cartographic data available for a study area may be a limiting factor in generating the four source-data layers required by Basinsoft. In general, Basinsoft can process most scales and sources of cartographic data available for a study area once the preprocessing is complete for the four source-data layers (Majure and Soenksen, 1991; Majure and Eash, 1991; Eash, 1993).

OPTIONAL PROGRAMS

Optional programs included in the tools directory (Figure 4.1) for use with Basinsoft are described below. The tools directory contains several variations of some programs, such as dem2grid.aml, which was developed to process one DEM at a time. The dem3grid.aml and dem4grid.aml process three and four DEM files at a time, respectively.

Area Weighting

A program to weight data by area, named areaweight.aml, is located in the tools directory. This program can be used as a module or as a stand-alone program to quantify characteristics from a variety of data, such as climatic data (annual precipitation and the two-year 24-hour precipitation intensity), which are stored in multipolygonal data layers (Eash, 1993). Minimal preprocessing is needed to execute this module. The areaweight.aml requires statewide or large area multipolygonal data layers representing the distribution of a characteristic, such as precipitation values, land-use values, pedologic values, geologic values, etc. The multipolygonal data layer must be larger than and encompass the drainage-divide data layer. Output is written to an auxiliary file specified by the user on the command line.

Dem2grid

The optional dem2grid.aml preprocessing program, located in the tools directory, is designed to convert a DEM file to a lattice data layer and project it into a user-specified projection. This module creates the required lattice data layer. There are several variations of this program in the tools directory. The various forms of dem2grid simultaneously preprocess up to four DEM files at a time.

Graphical Output

The graphical output module, graphics.aml, is located in the tools directory and can be used as a module or as a postprocessing program. Graphical output can be used in the interpretation of output generated by Basinsoft and can be useful in quality-control checking for obvious errors. The graphical output is in the form of an ARC/INFO graphics file which can be readily converted to a postscript file. Figure 4.5 (see color section) shows an example of graphical output. Data contained in the graphical output include: (1) tables of the basin characteristics quantified, (2) variables used by Basinsoft, and (3) a plot of the drainage-divide data layer, elevation contours, main channel, basin-length measurement line, points depicting the 10% and 85% distances along the main channel from the basin outlet, and a point at the basin outlet (outfall).

CONCLUSIONS

A computer program named Basinsoft has been developed by the U.S. Geological Survey to quantify 27 morphometric drainage basin characteristics using a geographic information system. Basinsoft was developed entirely using AML to increase portability among systems. Basinsoft uses ARC/INFO AMLs written for ARC/INFO version 7.0.2 or later. Basinsoft requires four source-data layers representing the drainage-divide, hydrography, hypsography, and a lattice elevation model of a selected drainage basin. Minimal preprocessing is required to prepare the source data used by Basinsoft. Optional programs and modules were developed to enhance the usability and functionality of Basinsoft.

Statistical comparison tests indicate that Basinsoft quantifications are not significantly different (p-value >0.05) from manual topographic-map measurements for nine of ten basin characteristics tested. Results also indicate that elevation contours generated by ARC/INFO from 1:250,000-scale DEM data are substantially overgeneralized when compared to elevation contours shown on 1:250,000-scale topographic maps and that quantification of basin slope thus is underestimated using the DEM data. A comparison test indicated that the Basinsoft module used to quantify basin slope is valid and that differences in quantified basin slope are due to source-data differences.

Basinsoft provides an automated computer procedure for the quantification of drainage basin characteristics and reduces the amount of time required to quantify drainage basin characteristics when compared to manual methods of measurement.

ACKNOWLEDGMENTS

The authors express their gratitude to the Spatial Data Support Unit (SDSU), Office of the Assistant Chief Hydrologist for Technical Support, USGS, for providing support in the development of Basinsoft; especially to Bob Pierce, Chief, SDSU, for providing technical support to the project, and to Mark Negri, Computer Specialist, SDSU, for providing programming expertise during the development of Basinsoft.

REFERENCES

Dempster, G. R., 1983. Instructions for Streamflow/Basin Characteristics File. U.S. Geological Survey National Water Data Storage and Retrieval System (WATSTORE), Vol. 4, Chap. II, sect. A.

Eash, D. A., 1993. Estimating Design-Flood Discharges for Streams in Iowa Using Drainage-Basin and Channel-Geometry Characteristics. U.S. Geological Survey Water-Resources Investigations Report 93–4062.

Eash, D. A., 1994. A Geographic Information System Procedure to Quantify Drainage-Basin Characteristics. *Water Resources Bulletin*, 30(1):1–8.

Environmental Systems Research Institute, Inc. (ESRI), 1994. ARC/INFO Users Guide, Version 7.0. Redlands, CA.

Harvey, C. A., and D. A. Eash, 1996. Description, Instructions, and Verification for Basinsoft, a Computer Program to Quantify Drainage-Basin Characteristics. U.S. Geological Survey Water-Resources Investigations Report 95–4287.

Jenson, S. K., and J. O. Dominique, 1988. Extracting topographic structure from digital elevation data for geographic information system analysis. *Photogrammetric Engineering and Remote Sensing,* 54, (11):1593–1600.

Majure, J. J., and D.A. Eash, 1991. An automated method to quantify physical basin characteristics. In *U.S. Geological Survey Toxic Substances Hydrology Program*, G. E. Mallard and D. A. Aronson, Eds. Proceedings of the Technical Meeting, Monterey, CA, March 11–15, 1991. U.S. Geological Survey Water-Resources Investigations Report 91–4034, pp. 558–561.

Majure, J. J., and P. J. Soenksen, 1991. Using a geographic information system to determine physical basin characteristics for use in flood-frequency equations. In *U.S. Geological Survey National Computer Technology Meeting*, B. H. Balthrop and J. E. Terry (Eds.) Proceedings, Phoenix, AZ, November 14–18, 1988. U.S. Geological Survey Water-Resources Investigations Report 90–4162, pp. 31–40.

Melton, M. A., 1957. An Analysis of the Relations Among Elements of Climate, Surface Properties, and Geomorphology. Office of Naval Research, Geography Branch, Columbia Univ. Dept. of Geology, New York, Technical Report 11, Project NR 389–042.

Melton, M. A., 1958. Geometric properties of mature drainage systems and their representation in an E_4 phase space. *Journal of Geology*, 66:35–54.

Office of Water Data Coordination, 1978. Physical basin characteristics for hydrologic analysis. In *National Handbook of Recommended Methods for Water-Data Acquisition*. U.S. Geological Survey, Reston, VA., Chap. 7.

Robbins, C. H., 1986. Techniques for Simulating Flood Hydrographs and Estimating Flood Volumes for Ungaged Basins in Central Tennessee. U.S. Geological Survey Water-Resources Investigations Report 86–4192.

Strahler, A. N., 1952. Hypsometric (area-altitude) analysis of erosional topography. *Bulletin of the Geological Society of America*, 63:1117–1142.

Strahler, A. N., 1958. Dimensional analysis applied to fluvially eroded landforms. *Bulletin of the Geological Society of America*, 69:279–300.

CHAPTER 5

Deriving Stream Channel Morphology Using GIS-Based Watershed Analysis

Scott N. Miller, D. Phillip Guertin, and David C. Goodrich

INTRODUCTION

Because of the time and degree of technical skill required for the completion of geomorphology studies, individual projects have historically been limited in size and scope. With the advent of geographic information systems (GIS), these technical problems have been assuaged. The GIS capability of storing large and diverse quantities of spatial data allows for the complex analysis of many sites to be carried out quickly, efficiently, and with a high degree of repeatability (Burrough, 1986). However, GIS-based projects often fail to integrate field-collected data with GIS spatial data. This project was designed to relate the GIS characterization of spatially distributed watershed characteristics with field measurements of point-attribute data (channel cross-section surveys). These data sets were related using statistical analysis to derive relationships between watershed characteristics and channel shape.

Watershed characterization based on geometric and physical properties was carried out in a GIS on 222 subwatersheds within the Walnut Gulch Experimental Watershed. At the same time a field measurement program was completed in which channel shape characteristics were measured at the outlet of each subwatershed. Statistical analysis between the two data sets showed a strong relationship between channel shape and watershed characteristics. It was shown that the derivation of hydrologic model parameters may be effectively carried out in a GIS on a large number of data points in a relatively short amount of time.

With its long history of data collection and observational data, Walnut Gulch serves as an excellent location on which to conduct research into geomorphologic and hydrologic processes (Renard et al., 1993). Relatively little work, however, has focused on the characterization of the entire watershed. This lack of data has limited the ability to model landscape processes on a basin scale (Lane et al., 1994). Additionally, most of the research on the relationship between channel and watershed characteristics has been conducted on intermittent and perennial streams. A knowledge gap therefore exists for this type of data on aridland watersheds and the processes acting on ephemeral channel systems such as exist on Walnut Gulch (Osterkamp et al., 1983; Lane et al., 1994). Instead of limiting this work to a small section of the watershed, it was decided to characterize as much of the watershed as possible. Sample sites were located randomly across the entire 148 km^2 watershed within all soil types and many hydrologic conditions. Strahler ordering analysis (Strahler, 1952) and other measures of channel and watershed characteristics were utilized to describe the watershed as quantitatively and thoroughly as possible.

Analyses of basin characteristics have been carried out in a GIS environment for many years (Burrough, 1986; Garbrecht and Martz, 1995), but many of these processes were found to be incompatible or unworkable for the data collected during this project. Therefore, a suite of GIS analysis tools and ARC/INFO Macro Language (AML) programs was developed to facilitate the GIS investigation (trade names are mentioned solely for the purpose of providing specific information and do not imply recommendation or endorsement by the U.S. Department of Agriculture). Channel shape, required for hydraulic routing, cannot be accurately predicted (or extracted) from DEMs. Therefore a principal goal of this project was to develop a methodology for predicting channel shape from watershed characteristics that could be readily derived from commonly available GIS coverages. During this process, field research was synthesized with GIS applications and photogrammetry to more thoroughly describe the channel and geomorphologic characteristics than had previously been attempted.

SITE DESCRIPTION OF WALNUT GULCH

Located in southeastern Arizona (approx. 110°W, 31°45′N) and comprised of rolling hills and some steep terrain, the elevation of Walnut Gulch Experimental Watershed ranges between 1190 and 2150 m A.M.S.L. Some urbanization exists in and around the town of Tombstone, but cattle grazing and recreational activities are the major land uses. Vegetation within the watershed is representative of the transition zone between the Chihuahua and Sonoran deserts, and is composed primarily of grassland and shrub-steppe rangeland vegetation.

Underlying Walnut Gulch is the geology of a high alluvial fan contributing to the San Pedro River watershed (Renard et al., 1993). Due to the enormous thickness and extent of the alluvial fill, the groundwater reserves are substantial, and can be found at depths ranging from 50 to 145 m (Libby et al., 1970). Some geologic control over the hydrology exists in the western regions of the watershed where metamorphic and orogenic activity has resulted in the fracturing and faulting of the bedrock. In 1994 the USDA Soil Conservation Service completed a detailed soil survey, finding that the watershed is dominated by 30 principal soil types (Breckenfield et al., 1995). Major soil units are Elgin-Stronghold (*Ustollic Paleargid, Ustollic Calciorthid*), Luckyhills-McNeal (*Ustochreptic Calciorthid*), McAllister-Stronghold (*Ustollic Haplargid, Ustollic Calciorthid*), and Tombstone (*Ustollic Calciorthid*).

The climate of Walnut Gulch can be classified as semiarid or steppe. Mean annual temperature in the city of Tombstone is 17.6°C, with a mean annual precipitation of 324 mm. Annual precipitation is highly variable in both timing and total depth. Rain occurs mainly during two seasons: summer rains are the product of monsoonal, highly localized, convective storms; winter rains are generally low-intensity events that cover a larger proportion of the watershed. The majority of runoff occurring on Walnut Gulch is the product of summer storms, and is therefore episodic and of relatively high intensity (Renard et al., 1993).

FIELD DATA COLLECTION

A field measurement program was undertaken wherein 222 channel cross-sections were surveyed for morphometric assessment. To account for basin-scale variability, a large number of randomly selected points were used, and multiple measurements were taken at each site to account for local variability in channel shape. These randomly located sample point locations were prestratified by soil type using a GIS procedure: each major soil type was assigned a weighted proportion of the sample points based on the areal extent of the soil coverage. At each site three cross-section surveys were taken to characterize the channel section just above the outlet of the subwatershed. Width and depth were measured at breakpoints (changes in slope). The three surveys were then combined to determine the average width and depth for the channel segment, and these results were combined to derive average cross-sectional area.

A strict protocol was followed at each sample location in order to ensure proper measurement and consistency between sites. Upon arrival at a site, an inspection of the bank morphology, vegetation, and soil characteristics along the entire reach was completed to ensure that cross-sections were located where they would be most representative of the channel section. A site description was recorded in a logbook for future analysis, and potential problems related to channel complexity and morphology were noted where applicable. Bankfull indicators, including slope breaks, changes in bed or bank materials, a shift in vegetative type, debris lines, and bank staining were noted in order to determine the bankfull depth (Osterkamp et al., 1983; Gordon et al., 1992; Harrelson et al., 1994). Wherever possible, evidence indicative of a constructive, rather than destructive process, was used to determine bankfull depth. In the southwestern United States, channel processes are governed by rapid and violent runoff events, and many of the channels on Walnut Gulch are actively degrading. Channels that were clearly degrading and out of equilibrium were not subjected to channel measurement since an adequate determination of bankfull depth was not possible.

At each of the site locations a minimum of three cross-sections was surveyed. If the channel reach was complex, up to five cross-sections were measured to ensure adequate representation. At each of the cross-sections a light line was pulled level across the channel top at the bankfull depth. The line was leveled and pulled taut to reduce sag. Measurements of channel depth and distance from the left bank (looking upstream) were taken at each break in slope across the cross-section.

Channel width was more easily measured with precision than channel depth. Although determining the stage to which floodwaters rise proved difficult, the possibility for error was greater when measuring depth. This is due to a number of factors. First, depth was only measured at break points, which are to some degree subjective. Second, there was always a slight amount of sag in the line when it was stretched across a channel, lending a source of imprecision to the depth measurements. Third, more random deposition or scour of the stream channels tends to impact local channel depth measurements to a greater degree than width measurements.

GIS DATABASE DEVELOPMENT AND ANALYSIS

Given Walnut Gulch's history as a research site into various aspects of hydrology and natural resource management and its extensive rainfall and runoff database, it was decided that the GIS database would be created at a resolution not ordinarily attempted. Throughout the database development, an answer to a basic question was sought: what are the highest levels of precision and accuracy that could be achieved? There can be a tendency by GIS developers to overestimate the level to which data may be discretized. By attempting to create maps with a higher resolution than is allowable by the data, errors may be introduced, and a false level of analysis can be attempted (Burrough, 1986). Fortunately, data available for Walnut Gulch were of a quality that allowed for a very high level of resolution and positional accuracy.

Of particular relevance was the stream channel coverage. In many GIS studies, the channel network is derived from a DEM in a raster environment and then translated into vector data. Alternatively, channels may be digitized from USGS topographic maps, but channels drawn on these maps are partly based on DEMs. Traditional GIS technique dictates that the majority of channels be digitized as single vectors bisecting the channel position, with a few of the larger drainages characterized with two lines to illustrate relative width. Since a correlation was to be made in this study between channel shape and watershed variables, a channel network map was constructed whereby only the smallest channels were digitized as single vectors. Channels wider than approximately 1.5 meters were drawn as polygonal features. This highly detailed procedure relied on the 1:5,000 orthophotographs as the base from which the stream positions and characteristics could be extracted. Most of the channels on Walnut Gulch were thus characterized in the GIS database as polygons, with associated width characteristics. In addition, where channel islands and bars were visible on the orthophotographs they were digitized. Thus, the channel network theme layer provides a detailed record of the channel system and its hydrologic characteristics as existed at the time the aerial photographs were taken (April 1988).

An important variable for the understanding of geomorphologic relationships is stream order. In this case the intensive channel network map was a drawback: because most of the channels were digitized as polygonal features it was not possible to automatically order the streams. To take advantage of GIS arc-node topology, the stream channels were vectorized. First, the map was translated into GRASS and rasterized with a one-meter resolution. The GRASS module "r.thin", which draws a parallel bisector through polygons, was executed on the stream map (Geographic Resources Analysis Support System, 1991). Upon completion of the vectorizing process, the maps were appended together and edited to remove spurious vectors created as a by-product of the thinning process.

The vector stream channel map was then re-exported into ARC/INFO, which supports both vector and routing functions. An ordering routine was created that takes advantage of the "from" and "to" node data structure that ARC/INFO imposes on vector maps. All the streams first had to be oriented in the downward direction (i.e., pointing downstream). Once the streams were all pointing in the downstream direction, the ordering program was initiated. By assigning all vectors that had an open-ended "from" node an order value of one, it was possible to stimulate a cascading effect, whereby all vectors were assigned a stream order based on their relationship and connectivity to other channels.

A 10 m resolution DEM provided the basis for the articulation of subwatersheds and the creation of many theme layers important to the statistical analysis of field data. Created from a large number of spot elevation points, contour data, and a thinned version of the channel network using the ARC/INFO tool "topogrid" (Environmental Systems Research Institute, 1994), the DEM was resolved to a 10 by 10 meter gridded surface. Using the "selectpoint" option within the "watershed" command in GRID, subwatersheds were delineated above each of the 222 channels surveyed in the field. From the DEM theme layers for flow direction and flow length were created for each watershed. Watershed characteristics that were derived with the GIS included: watershed area; maximum flow length; average slope; elevation characteristics; and watershed shape variables.

STATISTICAL ANALYSIS

Descriptive statistics implied that stream order was significantly related to channel shape variables. An analysis of variance showed that significant differences exist for channel width, depth, and cross-sectional area between each stream order. Multiple analysis of variance proved average watershed soil clay content to have no influence on channel shape. Relationships between channel shape variables and watershed parameters were investigated using simple linear regression analysis. Having found strong relationships between these variable sets, multiple regression analysis was employed to further refine these relationships.

RESULTS

For the purposes of evaluating the relationship between channel morphology and the contributing area, the relationships describing the channel cross-sectional area were of primary interest, and

Table 5.1. Relationship of Channel Morphology Variables to Stream Order

Order / N	Average Width (cm)	Average Depth (cm)	Average Cross-Sectional Area (m²)
1 / 58	279.65	26.32	0.802
2 / 65	404.32	34.57	1.47
3 / 40	563.03	40.10	2.54
4 / 26	960.39	54.94	5.63
5 / 20	1967.42	52.58	10.58
6 / 13	3329.99	79.69	26.21

deterministic models were derived using regression analysis. Channel cross-sectional area is a function of both channel width and average depth and thus reflects the total channel response to its hydrologic regime. Channel width can be extracted from a high resolution GIS such as exists for Walnut Gulch. Therefore, given a strong statistical relationship between cross-sectional area and watershed parameters, it would be possible to fully articulate channel geometry (width, depth, cross-sectional area) for all channels throughout Walnut Gulch. This ability to model channel shape accurately when a minimum of field data are available may benefit the application of a host of hydrologic models that incorporate hydraulic channel routing (i.e., the USGS DR3M model— Alley and Smith, 1982; KINEROS—Woolhiser et al., 1990; HEC–1, Army Corps of Engineers— Feldman, 1995).

Horton (1945) investigated the role of stream order on channel shape and hydrologic processes. He found that stream order was highly correlated to many watershed and channel variables, and that stream order could be used as a predictive tool for these variables. Strong relationships between stream order and channel shape were also found to exist on Walnut Gulch (Table 5.1). In this project, statistically significant differences were found to exist between the means of channel width, depth, and cross-sectional area for each step in stream order. Stream order, which is closely related to contributing area, was found to exert a strong effect on channel shape, and was used to stratify the data into subcategories for further analysis.

Channel characteristics were related strongly and in a semilog fashion to stream order. Average channel width, depth, and cross-sectional area were all directly related to order, with a break in the trend occurring between the fourth and fifth order channels, but only for channel width and depth; cross-sectional area maintains a semilog relationship throughout each step in order. The average value for channel depth shows a decrease between channel orders four and five, which is out of trend for every other increase in order (Figure 5.1). However, there is a significant increase in channel width between the fourth and fifth order channels, effectively counteracting the decrease in depth so that the relationship between cross-sectional area and stream order remains consistent

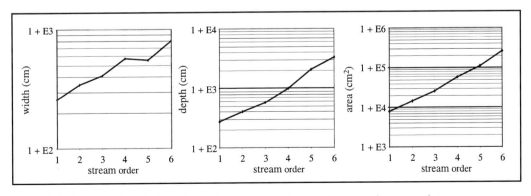

Figure 5.1. Semi-log plots of channel morphologic properties as a function of stream order.

across each order. The overall effect on channel shape is an increase in the channel width:depth ratio, while the relationship of cross-sectional area to order (and, hence, upstream watershed area) remains consistent (Figure 5.1).

Statistical Properties of Channel Shape

Channel width appears to be more sensitive to the influence of watershed parameters than channel depth. Measured values of width have a large spread in their data, while the values for depth show a more central tendency with a lower variation. Without exception, channel width proved to have a higher coefficient of determination than depth (e.g., $R^2 = 0.33$ for depth and 0.72 for width when related on a log-log basis to watershed area) when regression analysis was performed. In fact, depth proved to be resistant to any deterministic model based on the variables used in this study. Some of this resistance to forming a deterministic relationship may be a function of the difficulties associated with precisely measuring depth in the field. Fluvial characteristics are undoubtedly important to this tendency: as flow energy increases in a channel, the channel will adjust its shape to accommodate the increased level of power and erosive energy. This can be accomplished through the widening and/or deepening of the channel. In the loosely consolidated soils of Walnut Gulch, the channels appear to respond to elevated flow energy by increasing their channel width proportionally more than depth.

Responding to the runoff they receive from uplands, stream channels constantly adjust their shape to achieve equilibrium with the flow volume. Changes in channel morphology may result in either degradation or aggradation, with a resultant change in the width:depth ratio, but the net effect is a change in the channel cross-sectional area. As such, the measurement and analysis of channel cross-sectional area is an effective method of illustrating the manner in which channels are responding to watershed characteristics.

A strong relationship exists between channel area and the maximum flow length within a watershed ($R^2 = 0.79$). Table 5.2 shows the results of regression models involving channel area. Long flow lengths within a watershed have been directly related to discharge (Leopold et al., 1964). With higher flows, the channel will become enlarged, either through bed scour or bank erosion, to accommodate the larger flows, resulting in an increased channel cross-sectional area. Following the same reasoning, a strong relationship between channel cross-sectional area and watershed area would also be expected. Data collected in this research support that logical extension. A log-log relationship ($R^2 = 0.68$) exists between channel cross-sectional area and watershed area. A strong relationship ($R^2 = 0.77$) exists between channel cross-sectional area and the watershed area:perimeter ratio, a measurement of the rotundity of a basin, and hence an indicator of basin response. Neither average watershed slope nor the relief ratio correlated strongly with channel cross-sectional area. The log of cumulative drainage length (total length of all channels in a subwatershed) had a moderate relationship to the log of channel cross-sectional area ($R^2 = 0.62$).

In order to improve on the relationships derived using simple linear regression, channel variables were related to watershed characteristics using multiple linear regression. Multiple regres-

Table 5.2. Results of Linear Regression Analysis between Channel Area and Watershed Variables

Variable	Watershed Characteristic	R^2	Coefficient	Constant	Se_{yx}
Channel area	maximum flow length	0.79	0.001	1.83	3.46
Log channel area	log watershed area	0.68	0.49	−2.44	0.34
Channel area	area:perimeter ratio	0.77	0.03	0.17	3.60
Log channel area	log cumulative channel length	0.62	0.51	−1.38	0.40

Table 5.3. Results of Stepwise Backwards Multiple Linear Regression Analysis for Channel Cross-Sectional Area as a Function of Watershed Variables[a]

Case	Regression Model	R^2	Se_{yx}
1	Ca = 0.686(So) + 0.065(Aw) + 0.909(Lm) − 0.006(h)	0.849	3.36
2	Ca = 0.40(So) + 0.009(Aw) + 0.821(Lm) − 0.006(h)	0.851	3.35
3	Ca = 0.72(So) + 0.095(Aw) +0.001(Lm) − 0.007(h) − 0.001(Dl)	0.851	3.34
4	Ca = 0.616(So) + 0.001(Lm) + 0.001(S)	0.849	3.42

[a] where: Ca = channel cross-sectional area (m^2); So = stream order; Aw = subwatershed area (m^2); Lm = maximum flow length (m); h = relief (m); Dl = sum of drainage lengths (m); S = basin slope.

sion analysis of channel cross-sectional area revealed the relatively strong role that channel order played in the determination of channel cross-sectional area. Systematic exploration of the watershed data, using both stepwise forward and backward regression analysis, showed that channel area was heavily dependent on stream order and the area of and maximum flow length within the contributing watershed (Table 5.3). Depending on the subset of parameters investigated, it was possible to extract a significant regression model with a number of different independent variables. To avoid collinearity, multiple pools of data were used during the regression analysis. For example, the relief ratio, a product of the maximum flow length and maximum elevation change, was considered separately from those two variables. The same separation was used for basin shape variables and watershed size. Note that a constant was not used in the analysis, and the equations were driven through the origin.

CONCLUSIONS

Strong statistical relationships were derived between channel variables measured in the field, such as width, depth, and cross-sectional area, and a host of watershed parameters, including channel order, watershed area, shape, drainage properties, and elevation characteristics that were defined using a GIS. Channel cross-sectional area was related in a deterministic manner to multiple watershed variables, yielding models with strong coefficients of determination ($R^2 > 0.84$). Channel shape (and, hence, bankfull stage) may thus be predicted from watershed characteristics readily extracted from common GIS coverages.

Field data were successfully integrated with GIS-derived results. Channel cross-sectional area and other field-measured channel morphometric parameters were found to be strongly related to watershed characteristics extracted from a high-resolution GIS. It is preferable to collect field data when developing parameters for application in hydraulic routing models, but field collection can be costly and time consuming. The channel coverage created for Walnut Gulch contains information on channel width. Using the values for width that can be extracted for the GIS, in conjunction with the developed regression models, values for channel depth and cross-sectional area may be calculated for all channel segments within the watershed. Relationships developed upon verification outside Walnut Gulch have the potential to overcome the inability of DEMs to parameterize channel cross-section properties. In this fashion hydrologic models can be parameterized using a GIS to aid in the understanding of hydrologic processes in the southwestern United States.

ACKNOWLEDGMENTS

Support for this project was provided by the United States Department of Agriculture (USDA) Agricultural Research Service, Southwest Watershed Research Center, Tucson, Arizona. The authors would also like to thank the Advanced Resource Technology Group (ART), University of Arizona, for the use of equipment, space, and technical advice.

REFERENCES

Alley, W. M., and P. E. Smith, 1982. Distributed Routing Rainfall-Runoff Model—Version II. Computer Program Documentation User's Manual. USGS-WRD open-file report 82–344. Gulf Coast Hydroscience Center, NSTL Station, MS.

Breckenfield, D. J., W. A. Svetlik, and C. E. McGuire, 1995. Soil Survey of Walnut Gulch Experimental Watershed. United States Department of Agriculture, Soil Conservation Service.

Burrough, P. A., 1986. *Principles of Geographical Information Systems for Land Use Assessment.* Oxford University Press, New York, NY.

Environmental Systems Research Institute (ESRI), 1994. ARC/INFO ver. 7.0 manual (on-line documentation). Environmental Systems Research Institute Corp. Redlands, CA.

Feldman, A. D., 1995. HEC–1 flood hydrograph package. Chapter 4 of *Computer Models of Watershed Hydrology.* Water Resources Pub., Highlands Ranch, CO, pp. 119–150.

Garbrecht, J., and L. Martz, 1995. TOPAZ: An Automated Digital Landscape Analysis Tool for Topographic Evaluation, Drainage Identification, Watershed Segmentation, and Subcatchment Parameterization; TOPAZ Overview. USDA-ARS Publication NAWQL 95–1.

Geographic Resources Analysis Support System (GRASS), 1991. GRASS ver. 4.0 Manual. U.S. Army CERL. Champaign, IL.

Gordon, N. D., T. A. McMahon, and B. L. Finlayson, 1992. *Stream Hydrology An Introduction for Ecologists.* John Wiley and Sons, New York, NY.

Harrelson, C. C., C. L. Rawlins, and J. P. Potyondy, 1994. Stream Channel Reference Sites: An Illustrated Guide to Field Technique. USDA Forest Service General Technical Report RM–245.

Horton, R. E., 1945. Erosional development of streams and their drainage basins: Hydrophysical approach to quantitative morphology. *Geological Society of America Bulletin* 56:275–370.

Lane, L. J., M. H. Nichols, M. Hernandez, C. Manetsch, and W. R. Osterkamp, 1994. Variability in discharge, stream power, and particle-size distributions in ephemeral-stream channel systems. In *Variability in Stream Erosion and Sediment Transport; Proceedings of the Canberra Symposium, December 1994.* IAHS Publication 224: 335–342.

Leopold, L. B., M. G. Wolman, and J. P. Miller, 1964. *Fluvial Processes in Geomorphology.* W.H. Freeman and Co., San Francisco, CA.

Libby, F. J., D. E. Wallace, and D. P. Spangler, 1970. Seismic Refraction Studies of the Subsurface Geology of the Walnut Gulch Experimental Watershed, AZ, USDA-ARS 41–164.

Osterkamp, W.R., L.J. Lane, and G.R. Foster, 1983. An Analytical Treatment of Channel-Morphology Relations. USGS Professional Paper 1288.

Renard, K. G., L. J. Lane, J. R. Simanton, W. E. Emmerich, J. J. Stone, M. A. Weltz, D. C. Goodrich, and D. S. Yakowitz, 1993. Agricultural impacts in an arid environment: Walnut Gulch studies. *Hydrologic Science and Technology* 9(1–4): 145–190.

Strahler, A. N., 1952. Dynamic basis of geomorphology. *Geological Society of America Bulletin* 63:923–938.

Woolhiser, D. A., R. E. Smith, and D. C. Goodrich, 1990. KINEROS- A Kinematic Runoff and Erosion Model; Documentation and User Manual. USDA-ARS Pub. ARS–77.

CHAPTER 6

GIS Modeling and Visualization of the Water Balance during the 1993 Midwest Flood

Pawel J. Mizgalewicz, W. Scott White, David R. Maidment, and Merrill K. Ridd

INTRODUCTION

The characterization of the water balance during a major flooding event is useful to scientists, land-use planners, and government agencies who need such information to develop strategies for coping with future floods. Recent technological developments in the form of computer mapping and analysis databases have provided the means to accurately analyze flood-related data sets, such as streamflow and precipitation measurements, over extended periods of time. This study investigates the use of geographic information systems (GIS) to model spatially distributed and time-varying hydrologic and meteorologic data sets, and the use of scientific visualization techniques in the interpretation of the results. The data sets referred to deal specifically with the 1993 Midwest flood which affected a large part of the Upper Mississippi River Basin.

The study area encompasses the main stem of the Mississippi River above Cairo, Illinois, and the main stem of the Missouri River below Gavins Point Dam near Yankton, South Dakota. Also included are the many tributaries of these rivers such as the James, Des Moines, Illinois, and many other rivers and streams. This region of nearly 700,000 km^2 was previously defined by the Scientific Assessment and Strategy Team (SAST) as the area most affected by the spring and summer floods of 1993 (SAST, 1994). States included in this study are Illinois, Indiana, Iowa, Kansas, Minnesota, Missouri, Nebraska, North Dakota, South Dakota, and Wisconsin (Figure 6.1). Exactly 180 Hydrologic Unit Code (HUC) subbasins (8 digit) cover the study area.

The goal of this project is twofold:

1. Develop, demonstrate, and document procedures used to model the water balance of a large-scale flooding event, such as the 1993 Midwest floods, utilizing hydrologic and meteorologic data sets in a GIS database; and
2. Develop a comprehensive understanding, in terms meaningful to scientists as well as policy and decision makers, of the climatic and hydrologic conditions related to the 1993 floods and the spatiotemporal dynamics of the events. This understanding will result from the use of visualization software in the depiction of water storage change over the land surface during the flooding period.

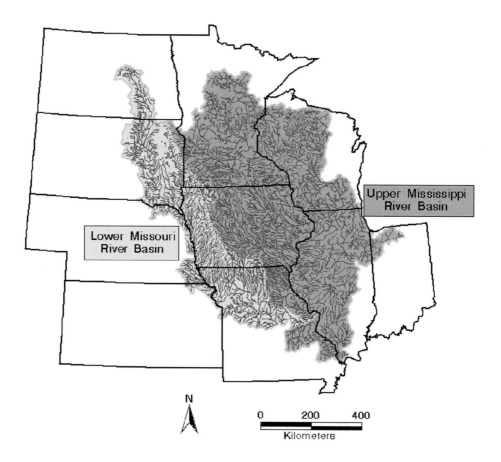

Figure 6.1. Study area—Upper Mississippi and Lower Missouri River Basins.

SURFACE HYDROLOGY CONSIDERATIONS

The water balance of a particular area represents a measure of the inputs to the hydrologic system and the outputs from that system over a specified period of time. In the case of this study, daily water balance maps were desired, therefore a decision was made to partition the study area into incremental drainage units based on the location of stream gauge stations. These drainage units or gauging station zones provided the basis on which areal interpolation of precipitation measurements were made. The outlets of the 180 HUC boundaries rarely corresponded with a stream gauge site, thus necessitating the subdivision of the study area into customized zones. The creation of the gauging station zones is shown in Figure 6.2. There are three possible zones: (1) no inflow; (2) one inflow; or (3) two or more inflows.

The equation for calculating the water balance of each gauging station zone is:

$$dS/dt = I - Q \qquad (1)$$

where S is the volume of water stored in each gauging station zone, t is the time index, and I and Q are inflow to the gauging station zone and outflow from the gauging station zone, respectively. The term on the right ($I - Q$) can be rewritten to account for the area of the gauging station zone:

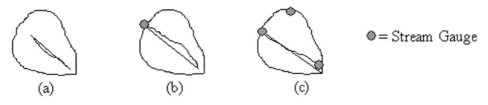

(a)   (b)   (c)   ◉= Stream Gauge

Figure 6.2. Gauging station zone configurations based on (a) no inflow, (b) one inflow, and (c) two or more inflows.

$$I - Q = (P - E)A + SQ_{in} - Q_{out} \tag{2}$$

where P is the rate of precipitation, E is the rate of evaporation, A is the area of the gauging station zone, SQ_{in} is the sum of the inflows recorded at gauges whose flows enter the zone and Q_{out} is the outflow at the downstream end of the zone.

The change in storage (ΔS) can be found by combining Equations 1 and 2 and integrating each term over a time interval of one day (Δt):

$$\Delta S = P\Delta tA - E\Delta tA + SQ_{in}\Delta t - Q_{out}\Delta t \tag{3}$$

In order to express ΔS in terms of average depth over the gauging station zone area, Equation 3 can be rewritten as the following ($\Delta y = \Delta S/A$):

$$\Delta y = P\Delta t - E\Delta t + (1/A)[SQ_{in}\Delta t - Q_{out}\Delta t] \tag{4}$$

where Δy is the change in storage depth per unit area of a gauging station zone. The values of Δy were computed on a daily basis (01/01/93 through 09/30/93) for each zone.

GIS AND WATERSHED DELINEATION

The raster GIS program GRID, a module available in the ARC/INFO® software package, allows for several hydrologic modeling procedures including the determination of flow direction, flow downstream accumulation, and watershed delineation. One of the strengths of this cell-based modeling package is the availability of map algebra functions. With map algebra, the variables in a logical expression actually are map (raster grid) layers. Algebraic manipulation of these grids can be performed at the local or individual cell level, neighborhood level (cells surrounding the cell of interest), zonal level (entire cell groups change in value), or at a global level (entire grid changes in value) (Tomlin, 1990).

Prior to delineating the gauging station zones, several data sets were acquired from the GCIP Reference Data Set (GREDS) CD-ROM, produced by the U.S. Geological Survey. This CD-ROM is a collection of several geographic reference data sets of interest to the global change research community (Rea and Cederstrand, 1994). ARC/INFO export files of 8-digit HUC boundaries and current streamflow gauge sites were obtained from this CD-ROM, as was a 15″ digital elevation model (DEM) of the region (500 meter resolution). RF1 river reach files in ARC/INFO export format for the Upper Mississippi and Missouri Rivers were downloaded via the Internet from the USGS node of the National Geospatial Data Clearinghouse (http://h2O.er.usgs.gov/nsdi/wais/water/rf1.HTML).

The first step in creating the gauging station zones was to edit the RF1 vector coverage to remove circular arcs and isolated lakes. Once a definite stream network was available, the vector coverage was converted to a raster grid, and then embedded into the 15″ DEM, which had been corrected for spurious sinks or pits. This embedding procedure "raised" the elevation of the off-stream grid cells surrounding the network in the DEM relative to the stream grid cells. By embedding these stream cells, a more precise network was created on the DEM that exactly matched the original paths of the RF1 vector coverage. This stream "burn in" technique has been shown to be particularly effective in areas of low relief (Maidment, 1996). The resulting grid was then clipped with a polygon coverage of the study boundaries. A 50 km buffer had been applied to this boundary coverage to account for drainage outside of the study area.

The next steps involved the actual delineation of the stream network within their embedded channels on the DEM. In ARC/INFO's version of this process, the flow direction is first determined by examining the neighborhood of eight grid cells that surround the cell of interest. The flow direction function identifies the lowest cell value in the neighborhood, and assigns a flow direction value to the corresponding cell in an output grid, thus creating an implicit stream network between cell centroids. Once this process was complete over the entire study area, a flow accumulation grid was made. The flow accumulation function in ARC/INFO uses the flow direction grid to determine the number of "upstream" cells. A new value is assigned to the cells in an output grid showing the number of cells that contribute or flow to downstream cells. High values indicate confluences of streams, whereas values of zero indicate watershed boundaries (Maidment, 1995). A conditional statement was then set up in ARC/INFO which isolated those cells that met a certain threshold of flow accumulation. Stream links were created, and the stream network was then in place.

After a suitable terrain model had been made, the next step was to precisely locate the USGS gauge stations on the stream grid. The point coverage of station locations was converted to a grid, and then viewed as a background grid against the stream grid. Unfortunately, most of the gauge cells did not lie directly on top of a stream cell, so the stream cell closest to a gauge cell was given a unique value (the USGS-assigned station number) to differentiate it from adjacent nongauge stream cells. Initially, 460 gauge cells were located on the stream grid; however, it was later determined that a number of these stations contained incomplete streamflow records. A total of 50 gauges were removed which did not have complete records. This left 410 gauges, which were unevenly distributed over the stream network. Some streams had many gauges, whereas other stream segments contained one or less. The gauge locations were viewed along with the HUC coverage to provide a better spatial representation of the gauges with respect to watersheds. It was decided that those gauges whose contributing drainage areas (an attribute in the ARC/INFO point coverage of the gauges) were less than 1,000 km^2 would be removed from the collection. Also removed were gauges that were concentrated in a particular watershed. Most HUCs contained between one and three gauge stations, but several contained more than three. Unless the HUC was large in area, those extra gauge stations exceeding three were also removed from the collection. Approximately 260 gauge stations on the stream grid were retained, and these gauge locations were for the most part uniformly distributed throughout the stream network.

The gauging station zones were then determined using the flow direction grid and the edited stream gauge grid (Figure 6.3). The resulting grid contained approximately 260 subbasins defined on the basis of the stream gauge locations. Each gauging station zone was checked against its corresponding HUC to ensure that the two sets of boundaries were mutually compatible.

GIS DATABASE AND MAP CREATION

Streamflow daily values for the 1993 water year were provided by the Water Resources Division of the USGS. After removing the 1992 values and reformatting the data into comma-delim-

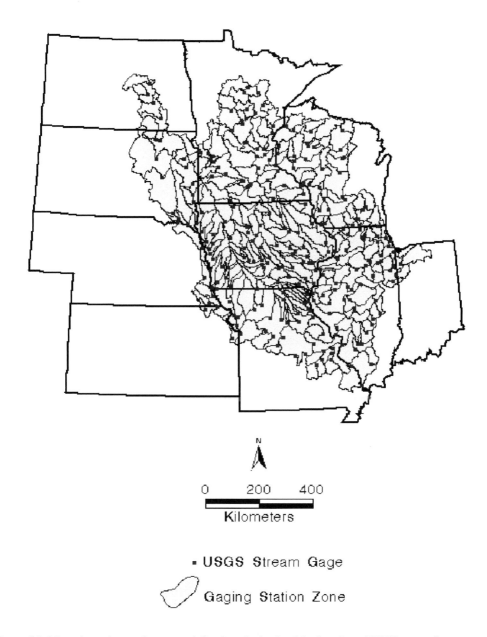

Figure 6.3. Map of gauging station zones defined on the basis of the location of USGS streamflow gauge stations.

ited ASCII text, the values were incorporated into the attribute table of the gauge station point coverage. This point coverage reflected the edits made to the stream gauge grid. Complete daily records for the gauge stations were assigned to the point coverage for a total of nearly 71,000 daily values from January 1 to September 30, 1993 (273 days).

Precipitation values obtained from National Climatic Data Center Summary of the Day files were treated in a manner similar to the streamflow data. A total of 1,078 climate stations were mapped. This number includes stations within the study area, and those within a 50 km buffer of the study area. Stations in the buffer zone were included so that a more precise interpolation of the precipitation measurements could be achieved for the regions at the study area borders. Over 290,000 precipitation measurements were incorporated into the point attribute table of the stations coverage. Some of the climate stations had missing precipitation measures, so an Arc Macro Language (AML) program was written to eliminate those stations with missing data.

A total of 273 daily precipitation grids with 4 km cell resolution was interpolated from the gauged precipitation data using the inverse distance weighting method. The ARC/INFO GRID function "zonalaverage" was subsequently used to obtain an average value of precipitation depth, as well as potential evapotranspiration (PET), over the gauging station zones. PET monthly values were compiled from the National Weather Service. Since values of PET vary slowly from month to month and the study area was saturated during the 1993 floods, it was felt that evaporation was most likely constant (i.e., one PET value per month) over the month in each gauging station zone.

The resulting grids show unique gauging station zones with average precipitation and evaporation depth values in each zone. Net streamflow grids show daily values averaged over each gauging station zone. These grids were created by subtracting Q_{out} from the sum of Q_{in}, and dividing the answer by the area of the gauging station zone. The final map series involve a combination of all of the preceding maps. Daily storage change (Δy) grids were produced using the map algebra functions in ARC/INFO GRID and Equation 4. Essentially, the equation changed from

$$\Delta y = P\Delta t - E\Delta t + (1/A)[SQ_{in}\Delta t - Q_{out}\Delta t] \, , \tag{4}$$

to

$$S \ Grid = P \ Grid - E \ Grid + (1/A)[NS \ Grid] \, , \tag{5}$$

where $S \ Grid$ is a grid of daily storage depth change, $P \ Grid$ is a daily precipitation grid, $E \ Grid$ is a daily evaporation grid, and $NS \ Grid$ is a daily net streamflow grid. Maps of cumulative storage depth were obtained by summing the grids. The resultant 273 grids show water storage depth of each gauging station zone in the study area on a daily basis.

Balancing the water inputs into the basin (precipitation and streamflow), and the outputs from the basin (evapotranspiration and runoff) meant that the database had to be extended to include the whole calendar year of 1993. The number of precipitation, streamflow, evaporation, and subsequent daily water storage grids was increased from 273 to 365 each to take into account the rest of the year (10/01/93 to 12/31/93). It was therefore assumed that the amount of water entering the basin on 01/01/93 was the same as the amount leaving the basin on 12/31/93, and an evaporation multiplication factor was introduced to make the necessary calculations. Based on these calculations, it was found that the water storage in the UMRB study area, when the basin is considered as one modeling unit, rose by approximately 110 mm in about a two-month time period (from the end of May to the beginning of August). However, the range of water storage depths was not geographically consistent throughout the basin. As one might expect, the highest storage depths increases took place in the most downstream units of the UMRB, where calculations show that the depth was nearly 300 mm for the two-month period described above.

VISUALIZATION

The final products of this study consisted of maps that represented the water storage depth which occurred on each subbasin or gauging station zone as a result of the heavy precipitation events between 01/01/93 and 12/31/93. Each day's maps represented a snapshot of how much precipitation occurred in the study area, and how it affected the land surface. Although the GIS can perform many of the tasks necessary to determine water storage, a final step remains in order to fully understand the temporal aspects of the flooding events. This visualization step involves the use of various computer software packages, coloring and animation procedures, and output types.

GIS software packages have provided the spatial sciences with an excellent set of tools for performing hydrologic analyses. The main problem with the technology today is its relative inability to handle temporal information. This chapter has described a method of incorporating temporal data into the attribute files of GIS layers, but the final product, as described up to this point, remains a series of static portrayals of a dynamic event. To overcome this problem, various visualization methodologies are being investigated as additional components to the GIS-based hydrologic modeling process. Buttenfield and Mackaness (1991) described GIS-related visualization as the interface of three processes: (1) computer analysis (data collection, organization, modeling, and representation); (2) human cognition (perception, pattern identification, and mental imaging); and (3) graphic design principles (construction of visual displays). MacEachren (1994) further described the concept of "geographic visualization" as stressing map use that can be conceptualized as a three-dimensional space. Both descriptions of visualization imply that maps and associated images can now be constructed that incorporate 3-D and 4-D perspectives. This ideally suits geographic visualization techniques as an end product to temporal representation using a GIS.

Currently, the authors are investigating several methods of geographic visualization. Most involve some form of geographic animation as a way of representing the temporal aspects of the flood. Initially, the 273 maps of water storage depth, averaged over each gauging station zone from 01/01/93 to 09/30/93, will be color-coded based on the depth of water storage in each zone. Blue is the most likely color, and will range in shades from light or pale (no water saturation of the land surface) to dark or bright (saturated land surface). These choropleth maps will be chronologically arranged from 01/01/93 to 09/30/93, and then put into an animation software package. Viewing these maps in sequence will provide the user with a dynamic display of water leaving gauging station zones and saturating downstream zones. Figure 6.4 (see color section) shows a sequence of four daily maps (07/29/93–08/01/93) with corresponding water storage in 8-digit HUC boundaries.

Another means of visualizing flow through the basin would be through the use of the stream network. The representation of the network in the GIS does not really account for its changing form and velocity during the flood months of 1993. A series of 273 numbers representing streamflow velocity could be assigned to the cells representing stream gauge locations along the rasterized network (500 m resolution), but the actual channel dimensions remain the same throughout the time period. Visualization procedures can be used in conjunction with the streamflow values to produce a dynamic representation of streamflow velocity through the basins. The flow velocity can be represented by some type of hydrologically significant symbol which appears to speed up or slow down depending upon the streamflow value at the gauge location. This iconic representation of streamflow will increase in size as flow becomes faster downstream. Combining the symbology of flow velocity with the choropleth map described above is also feasible.

The authors are also experimenting with various means of stream channel visualization. As previously mentioned, the streams reside in the GIS as a 500 meter wide grid network. Of course, the streams vary in size in the Upper Mississippi and Missouri River basins from creeks a few meters wide, to the Mississippi River at Cairo, Illinois, which is nearly 1 km wide during normal flow conditions. Vector representation of the stream network would allow some flexibility in the appearance of the channels, and this method is currently being investigated with the GIS software packages ARC/INFO and ArcView. Satellite and radar imagery will also be incorporated into a visualization package to provide a more realistic view of the flooding events. By using imagery draped over a DEM of the study area, three-dimensional viewpoints can be achieved which provide even greater flexibility in terms of visualization and human cognition of the floods.

SUMMARY AND CONCLUSIONS

In this study, geographic information system software has been utilized as a first step to visualizing the water balance that was in place during the 1993 Midwest floods. A series of daily maps for 1993 are being produced that combine streamflow, precipitation, and evaporation measurements. This series of maps, while informative and meaningful to the group of people who understand GIS, needs to be made available to the public in a different form. The second step will be the development of a dynamic representation of this temporal data through the use of geographic visualization procedures. The final product of this research will be useful to many different people who study such natural hazards as flooding. Methodologies developed in this study can be used to investigate future floods, and will hopefully provide the end users with several different views of the 1993 Midwest flood which will enable them to make wise decisions regarding floodplain management and land use practices.

ACKNOWLEDGMENTS

This research has been supported by a grant from the U.S. Geological Survey to Drs. Maidment and Ridd. Thanks are also due to Dr. William Kirby, of the USGS Water Resources Division in Reston, VA, for supplying the authors with necessary data sets, and to Dr. Barbara Buttenfield for providing some ideas about geographic visualization.

REFERENCES

Buttenfield, B.P., and W.A. Mackaness, 1991. Visualization, in Maguire, D., M. Goodchild, and D.W. Rhind, Eds., *Geographical Information Systems: Principles and Practice*. Longman, London, pp. 427–443.

MacEachren, A.M., 1994. Visualization in modern cartography: Setting the agenda, in MacEachren, A.M., and D.R. Fraser-Taylor, Eds., *Visualization in Modern Cartography*. Pergamon, Oxford, UK pp. 1–12.

Maidment, D.R., 1995. *GIS and Hydrology: A Workshop of the 15th Annual ESRI User Conference*. Environmental Systems Research Institute, Redlands, CA.

Maidment, D.R., 1996. GIS and Hydrologic Modeling—An Assessment of Progress. Paper Presented at the Third International Conference on GIS and Environmental Modeling, Santa Fe, NM. National Center for Geographic Information and Analysis, Santa Barbara, CA, 1 CD-ROM.

Rea, A., and J.R. Cederstrand, 1994. GCIP Reference Data Set (GREDS): U.S. Geological Survey Open-File Report 94–388. USGS, Reston, VA, 1 CD-ROM.

SAST, 1994. Science for Floodplain Management into the 21st Century, Preliminary Report of the Scientific Assessment and Strategy Team Report of the Interagency Floodplain Management Review Committee to the Administration Floodplain Management Task Force. Washington, D.C.

Tomlin, C.D., 1990. *Geographic Information Systems and Cartographic Modeling*. Prentice Hall, Englewood Cliffs, NJ.

CHAPTER 7

Selection, Development, and Use of GIS Coverages for the Little Washita River Research Watershed

Patrick J. Starks, Jurgen D. Garbrecht, F.R. Schiebe, J.M. Salisbury, and D.A. Waits

INTRODUCTION

The Little Washita River Watershed (LWRW) in south central Oklahoma (Figure 7.1) is the largest and one of the longest-studied research watersheds operated by the United States Department of Agriculture's (USDA) Agricultural Research Service (ARS). The watershed drains an area of 611 km² and has been studied since 1961 for rainfall runoff, impact of flood control structures, water quality, sediment transport and best management practices. A series of Geographical Information Systems (GIS) raster coverages have been developed to support present and future ARS research on the LWRW and to complement the historical and current databases. The selection of the

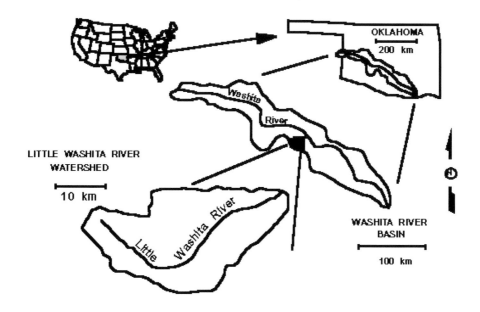

Figure 7.1. Location of the Little Washita River Watershed.

GIS coverages was guided by hydrologic research needs (Goodrich et al., 1994) and an effort to support both distributed and lumped parameter watershed modeling. The coverages are grouped into three categories: topography, soils, and land cover.

The object of this chapter is to present topics relating to the development and use of the three categories of GIS coverages. The selected topics address: (1) the development of hydrographic data layers from the Digital Elevation Model (DEM); (2) reliability of the soil property data extracted from the soils coverage and county soil survey data; and (3) land cover identification from Landsat satellite remotely sensed data.

DIGITAL ELEVATION MODEL AND TOPOGRAPHIC GIS DATA

Automated extraction of topographic parameters from DEMs has established itself in GIS over the past decade. This is attributed to the importance of and need for landscape derived data and to the increasing availability of DEMs and software products that derive topographic data from DEMs. In the field of water resources and hydrology, the main uses of digital landscapes are watershed segmentation, definition of drainage divides and channel networks, determination of catchment geometry, and parameterization of landscape properties such as terrain slope and aspect (Jenson and Domingue, 1988; Mark, 1988; Moore et al., 1991; Martz and Garbrecht, 1992). Such landscape evaluation tasks are generally tedious, time-consuming, error-prone, and often subjective when performed manually from topographic maps, field surveys, or photographic interpretations (Richards, 1981). The automated techniques are faster and provide more precise and reproducible measurements than traditional manual techniques applied to topographic maps (Tribe, 1991). Digital data generated by automated techniques also have the advantage that they can be readily imported and analyzed by GIS.

Most of the topographic coverages presented here were automatically derived from the DEM of the LWRW using software TOPAZ (Garbrecht and Martz, 1999). TOPAZ (TOpographic PArametriZation) is a software package for automated digital landscape analysis. It uses raster DEMs to identify and measure topographic features, define surface drainage, subdivide watersheds along drainage divides, quantify the drainage network, and parameterize subcatchments. TOPAZ is designed primarily for hydrologic and water resources investigations, but is equally applicable to address a variety of geomorphological, environmental and remote sensing applications. TOPAZ is discussed more fully elsewhere in this volume. Topographic GIS coverages of the LWRW, DEM resolution and quality, and degree of watershed segmentation and drainage density are addressed below.

Topographic GIS Coverages

A DEM with a horizontal and vertical resolution of 30×30 meters and 0.3 meter, respectively, is available for the LWRW. The DEM was developed in 1996 by the U.S. Geological Survey, Rolla, MO, from digitized contour data. The quality of the DEM corresponds to a Level 2 DEM; thus it has been processed at production time for consistency and edited to remove identifiable systematic errors. This DEM is the basic GIS coverage representing the landscape topography of the LWRW.

DEM preprocessing is necessary before deriving additional hydrographic and topographic data because DEMs commonly contain localized depressions and flat surfaces, many of which are artifacts of the horizontal and vertical DEM resolution, DEM generation method, and elevation data noise. Depressions and flat surfaces are problematic for drainage identifications. Depressions are sinks at the bottom of which drainage terminates, and flat surfaces have indeterminate drainage.

Therefore, TOPAZ preprocesses the input DEM to rectify these features and allows the unambiguous determination of the drainage over the entire digital landscape. Rectifications are strictly limited to cells of depressions and flat surfaces so as to minimize the impact on the overall information content of the elevation data. Further details on the rectification procedure are given below. With this rectified DEM the hydrographic and topographic coverages listed in Table 7.1 can be derived using software TOPAZ.

Table 7.1. Topographic GIS Coverages and Data Source

GIS Coverage	Source and Development Procedure
Digital Elevation Model	USGS, contour interpolation
Watershed boundary	TOPAZ, downslope flow routing concept
Drainage network	TOPAZ, downslope flow routing concept
Subcatchment drainage boundaries	TOPAZ, downslope flow routing concept
Terrain slope	TOPAZ, surface derivative
Terrain aspect	TOPAZ, surface derivative

The watershed boundary, subcatchment drainage divides, and drainage network computed by TOPAZ are based on the D8 method, the downslope flow routing concept, and the critical source area (CSA) concept. The D8 method (Douglas, 1986; Fairfield and Leymarie, 1991) defines the landscape properties for each individual raster cell as a function of itself and its eight immediately adjacent cells. The downslope flow routing concept defines the drainage on the landscape as the steepest downslope path from the cell of interest to one of its eight adjacent cells (Mark, 1984; O'Callaghan and Mark, 1984; Morris and Heerdegen, 1988). The CSA concept defines the channels draining the landscape as those raster cells that have an upstream drainage area greater than a threshold drainage area, called the critical source area (CSA). The CSA value defines a minimum drainage area above which a permanent channel is maintained (Mark, 1984; Martz and Garbrecht, 1992). The CSA concept controls the watershed segmentation and all resulting spatial and topologic drainage network and subcatchment characteristics.

DEM Resolution and Quality

The spatial resolution and quality of the DEM data are important considerations when deriving hydrographic data from DEMs. The choice of an appropriate resolution must be made under consideration of the landscape characteristics that are to be represented and the use of the derived data products. In the case of the LWRW it has previously been shown (Garbrecht and Martz, 1993) that the network and drainage divides can be adequately derived from a DEM with a 30×30 meter horizontal resolution.

A more difficult question was that of the quality of the elevation data. Accuracy standards, data noise, interpolation errors, and systematic production errors often create spurious depressions, flat areas, and flow blockages which cause problems for the identification of drainage features (Garbrecht and Starks, 1995). These in turn impact the drainage identifications and indirectly the drainage divides and network. The DEM for the LWRW is a Level 2 DEM and has been processed for consistency and systematic errors. However, a Level 2 DEM is not hydrographically corrected and problems such as spurious depression and flat areas remain. These must either be rectified, for example through data preprocessing as described above, or accepted as data limitations. For the LWRW project these problem features in the DEM have been rectified.

Spurious depressions are rectified by a procedure of depression breaching/filling (Garbrecht et al., 1996), and flat surfaces are rectified by imposing two gradients which are a function of the landscape configuration surrounding the flat surface (Garbrecht et al., 1996). Both these rectifications result in an approximated, yet realistic drainage pattern. It was also common to see discrepancies in the position of derived and actual channels. Such discrepancies were expected because a DEM is only an approximation of the true landscape topography, and the derived drainage divides and network represent the drainage features of the approximated topography.

Degree of Watershed Segmentation and Drainage Network Density

The degree of watershed segmentation and drainage network density are a function of the value given to the critical source area parameter (CSA). The CSA is the minimum drainage area that is required to initiate a first-order channel (Strahler, 1957). Any area smaller than the threshold value does not produce enough runoff to form and maintain a channel. The threshold CSA value depends upon, among other things, terrain slope characteristics, soil properties, land use, and climatic conditions. It can be as small as a fraction of a hectare or as large as tens of hectares, depending on the landscape characteristics under consideration.

Often the CSA value is also used to represent different degrees of watershed segmentation and drainage network densities to address scaling issues. At a small scale, one would choose a small CSA value to represent the smallest channels and hill slopes. As a result, a high degree of partitioning, many subcatchments, and a dense drainage network are obtained. In the case of large-scale applications, only the major streams in the watershed may be needed. A large CSA value would result in the desired low degree of watershed segmentation, few subcatchments, and only large streams. This capability to generate GIS coverages of different degrees of watershed segmentation and drainage network densities is important for landscape modeling.

Figure 7.2 (see color section) is an overlay of the basic LWRW DEM coverage and a TOPAZ-generated stream network from the DEM. The relief of the DEM is approximately 190 m. Each color change on the DEM represents a 7 m change in elevation from the neighboring color. In this example, a CSA of 8 hectares was used to generate the stream network, resulting in 1,218 first order, 264 second order, 54 third order, 12 fourth order, 4 fifth order, and 1 sixth order (Strahler number) channels.

DIGITAL SOILS COVERAGE

The LWRW soil coverages were developed from two data sources provided by the Natural Resources Conservation Service (NRCS): (1) county soil survey maps, and (2) the digital State Soil Geographic Database (STATSGO). The STATSGO database contains the soil attributes for the county soil survey maps. From this information a basic soil coverage and derived soil attribute coverages can be developed.

The basic soil coverage for the LWRW consists of the soil mapping units digitized from the county soil survey maps and rasterized to a 30 m pixel cell size. It was readily apparent that the name given to the soil mapping units is influenced by which county the mapping unit falls within (i.e., the county lines are seen in the soils coverage) [Figure 7.3 (see color section)]. This systematic trend is partly related to the different experience and interpretation of the county's soil scientist in charge of soil classification (Arnold et al., 1994). The trend can also be attributed to different dates at which a county's soils were classified. Soil names of an earlier classification in one county may not correspond to an updated soil classification in another county at a later date. However, it is not the difference in name that is important, but that the soil physical properties are similar for both soil names. Therefore, special care was taken to ensure that the attribute data were correct for all soils, regardless of soil name, before including them in the GIS data base for the LWRW. The consistency was ensured by involvement of NRCS personnel in the review of the soils attribute data, particularly for the soils along the county lines.

The soil attribute data were transferred from the NRCS STATSGO database into a computer worksheet. The data consist of both generalized attributes for the soil mapping units as well as soil profile data for up to six layers within a mapping unit. The soil layer data include surface and sub-surface horizons. Tables 7.2 and 7.3 list examples of attribute data for the soil mapping units and profile layers, respectively. Some attributes are given as maximum and minimum values, and it is up to the investigator's discretion as to which value within this range is appropriate for use in a particular application.

Table 7.2. Examples of Attribute Data for Soil Mapping Units

Slope	Surface Soil Texture
Taxonomic classification	Annual flooding frequency
Flood duration class	Depth to water table
Ponding depth of surface water	Depth to bedrock
Hydrologic soil group	

Table 7.3. Examples of Soil Profile Attribute Data

Depth of Soil Layer	Permeability
Soil texture	Clay content
Water holding capacity	Bulk density

A number of soil attribute coverages can be developed from the basic GIS soil coverage and the attribute data contained in the worksheet by matching the soil mapping units codes in the GIS coverages to soil characteristics in the worksheet attribute file and creating a corresponding soil attribute coverage.

LAND COVER

Land cover is a dynamic entity which varies both spatially and temporally. For example, in agricultural areas it is typical for crop canopies to cover a field for part of the year, while at other times that field is fallowed or bare. Also, crop rotation patterns, crop type, and total acreages planted in crops vary from year to year.

A series of land cover coverages was produced to gain a better understanding of the vegetative dynamics for the LWRW. These coverages were derived from Landsat MSS satellite images for the spring, summer, and fall seasons of every even-numbered year from 1972 through 1994. Each seasonal image was subjected to an unsupervised classification on all four wavebands from the satellite. In the unsupervised approach, a cluster analysis was used to examine the reflectance properties of the land surface and to aggregate related reflectance values into a number of classes. These classes were derived by cluster analysis and represent natural groupings of reflectance values (Eastman, 1992; Lillesand and Kiefer, 1994). Finally, ground truthing was used to assign a land cover category to each class. Land cover categories developed from the seasonal data are water, urban, bare soil, woodlands, native rangeland, tame rangeland (planted), and cultivated lands.

The seasonal remotely sensed data were combined to yield a synoptic temporal view of the agricultural landscape. This was achieved by converting the spring, summer, and fall remotely sensed data into "greenness" values using the greenness vegetation index algorithm (Kauth and Thomas, 1976). The greenness index quantifies vegetation presence and vigor. Colors were assigned to crops in accordance with their "greenness value" and in which season they were actively growing. For example, crops growing only in summer were assigned a color of green, with intensity of that color in proportion to the greenness value. By overlaying these three separate seasonal data sets, a new GIS coverage of the temporal dynamics of the agricultural landscape was produced [e.g., Figure 7.4 (see color section)]. Table 7.4 lists the GIS categories for the coverages of the temporal variability of land cover.

Table 7.4. GIS Categories for Temporal Land Cover Variability

Rangeland:	-native and planted
Wheat :	-spring only
	-fall only
	-spring only
Other crops:	-summer
	-spring and summer
Woodlands	
Water:	-also includes urban and bare soil

CONCLUSIONS

The development and use of three categories of GIS coverages for the LWRW were presented. Selected topics addressed the development of hydrographic coverages from a DEM, reliability of soil property data extracted from county soil survey data, and identification and dynamics of land cover derived from remotely sensed data. Hydrographic data sets were discussed with reference to the quality and spatial resolution of the DEM from which the coverages were derived, and also with reference to the degree of watershed segmentation and drainage network density. Development of the soils coverage revealed inconsistencies in the soil mapping units along county boundary lines within the LWRW. Therefore, NRCS personnel were involved in verifying the consistency of the soil attribute data before it was transferred from the STATSGO data files into the LWRW GIS. Two types of land cover information were derived from remotely sensed data. These show the spatial distribution and temporal dynamics of the land cover. Together these topographic, soil and land cover coverages provide a comprehensive GIS database in support of our research program in water resources and watershed modeling.

REFERENCES

Arnold, J.G., J.D. Garbrecht, and D.C. Goodrich, 1994. Geographic Information Systems and large area hydrology. In Richardson et al., Eds., *Proceedings of the ARS Conference on Hydrology*, Denver, CO, September 13–15, 1993. United States Department of Agriculture, Agricultural Research Service, 1994–1995.

Douglas, D. H., 1986. Experiments to locate ridges and channels to create a new type of digital elevation Model. *Cartographica*, 23(4):29–61.

Eastman, J.R., 1992. IDRISI: User's Guide. Clark University, Graduate School of Geography, Worcester, MA.

Fairfield, J., and P. Leymarie, 1991. Drainage Networks from Grid Digital Elevation Models. Water Resources Research, 27(4):29–61.

Garbrecht, J., and L.W. Martz, 1993. Network and subwatershed parameters extracted from digital elevation models: The Bills Creek experience. *Water Resources Bulletin*, 29(6):909–916.

Garbrecht J., and L. W. Martz, 1999. TOPAZ: An Automated Digital Landscape Analysis Tool for Topographic Evaluation, Drainage Identification, Watershed Segmentation and Subcatchment Parameterization: TOPAZ Overview. United States Department of Agriculture, Agricultural Research Service, Grazinglands Research Laboratory, El Reno, OK, ARS Pub. No. GRL 99-1, 26 pp.

Garbrecht, J., and P. J. Starks. 1995. Notes on the use of USGS Level 1 7.5 minute DEM coverages for landscape drainage analysis. *Photogrammetric Engineering and Remote Sensing*, 61(5):519–522.

Garbrecht, J., P. J. Starks, and L. W. Martz, 1996. New digital landscape parameterization methodologies. In Hallam et al., Eds., *Proceedings of the AWRA Annual Symposium, GIS and Water Resources*. American Water Resources Association Symposium, Herndon, VA, TPS–96–3.

Goodrich, D.C., P.J. Starks, R.R. Schnabel, and D.D. Bosch, 1994. Effective use of USDA-ARS experimental watersheds. In Richardson et al., Eds., *Proceedings of the ARS Conference on Hydrology*, Denver, CO, September 13–15, 1993. United States Department of Agriculture, Agricultural Research Service, 1994–1995.

Jenson, S. K., and J. O. Domingue, 1988. Extracting topographic structure from digital elevation data for Geographical Information System analysis. *Photogrammetric Engineering and Remote Sensing*, 54(11):1593–1600.

Kauth, R.J., and G.S. Thomas, 1976. The tasseled cap—A graphical description of spectral-temporal development of agricultural crops as seen by Landsat. *Proceedings of the Symposium on Machine Processing of Remotely Sensed Data*. Laboratory of Applied Remote Sensing, Purdue University, West Lafayette, IN, pp. 41–51.

Lillesand, T.M., and R.W. Kiefer, 1994. *Remote Sensing and Image Interpretation*, 3rd ed., John Wiley and Sons, New York, NY.

Mark, D. M., 1988. Network models in geomorphology. In Anderson, M. G., Ed., *Modeling Geomorphological Systems*. John Wiley and Sons, New York, NY, pp.73–96.

Mark, D. M., 1984. Automatic detection of drainage networks from digital elevation models. *Cartographica*, 21(2/3):168–178.

Martz, L.W., and J. Garbrecht, 1992. Numerical definition of drainage networks and subcatchment Areas from Digital Elevation Models. *Computers and Geosciences* 18(6):747–761.

Moore, I. D., R. B. Grayson, and A. R. Ladson, 1991. Digital terrain modeling: A review of hydrological, geomorphological and biological applications. *Hydrological Processes*, 5(1):3–30.

Morris, D. G., and R. G. Heerdegen, 1988. Automatically derived catchment boundary and channel networks and their hydrological applications. *Geomorphology*, 1(2):131–141.

O'Callaghan, J. K., and D. M. Mark, 1984. The extraction of drainage networks from digital elevation data. *Computer Vision, Graphics, and Image Processing*, 28:323–344.

Richards, N. S., 1981. General problems in morphology. In Gordie, A. S., Ed., *Geomorphological Techniques*. Allen and Umnin, London, UK, pp. 26–30.

Strahler, A.N., 1957. Quantitative analysis of watershed geomorphology. *Transactions of the American Geophysical Union*, 38(6):913–920.

Tribe, A., 1991. Automated recognition of valley heads from digital elevation models. *Earth Surface Processes and Landforms*, 16(1):33–49.

CHAPTER 8

Regional Characterization of Inland Valley Agroecosystems in West and Central Africa Using High-Resolution Remotely Sensed Data

Prasad S. Thenkabail and Christian Nolte

BACKGROUND

Inland valleys (IVs) (locally also known as fadamas, bas-fonds, and dambos) have the potential to become agroecosystems with a substantial impact on African food production (Izac et al., 1991). These agroecosystems are favorable for rice cultivation and dry season cropping and have the potential to increase acreage and yields in Africa if careful attention is paid to technical, environmental, and socioeconomic constraints (Juo and Lowe, 1986).

Inventorying, mapping and characterizing inland valleys at regional level (meso/semidetailed level, typically mapped on scales of 1:50,000 through 1:250,000) in sample areas of macrolevel agroecological zones (Figure 8.6, color section) is crucial for the selection of representative research sites, which then allows for the development of appropriate technologies that can be reliably tested and transferred to larger regions (regionalization or technology transfer).

Recognition of the importance of inland valley agroecosystems led the International Institute of Tropical Agriculture (IITA) to adopt a new research agenda, as put forth in Izac et al. (1991). This strategy involved a combination of biophysical and socioeconomic research issues to be addressed at three spatial scales:

a. level I (macro or continental or subcontinental, typically mapped in scales of greater than or equal to 1:5,000,000);

b. level II (meso or regional or semidetailed mapped in scales of 1:100,000 to 1:250,000); and

c. level III (micro or research site/watershed related, mapped in scales of less than or equal to 1:50,000).

This was conceived to facilitate the design of appropriate technology, able to sustain the highly varying resource base and at the same time to be acceptable to smallholder farmers in the diverse socioeconomic and ethnic environment of West and Central Africa.

The first step to characterize parameters critical to land use of IVs constituted the (macro) level I map based on secondary agroecological and soil data using a Geographic Information System (see Thenkabail and Nolte, 1995a for details). The objective was to map on a subcontinental

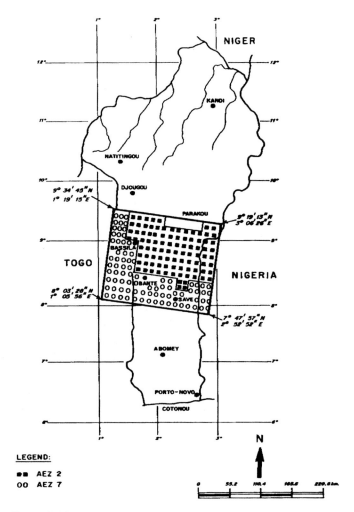

Figure 8.1 Spatial distribution of level 1 agroecological and soil zones (AEZ) in Landsat 192/54. Of a total area about 3.12 Mha, 49% fall into AEZ 2 and 46% into AEZ 7. The remaining 5% area is outside the 18 zones mapped in Figure 8.1.

(macro) scale broad agroecological and soil zones in the mandate area of IITA in West and Central Africa (Figure 8.1b). The level I map was the result of combining two parameters:

    a. IITA's agroecological zones; and

    b. major soil groupings according to the FAO classification (FAO/UNESCO 1974, 1977).

The five agroecological zones of IITA's map, namely northern Guinea savanna, southern Guinea savanna, derived and coastal savanna, humid forest, and midaltitude savanna, were overlaid with the 23 zones of major soil groupings. This resulted in 18 zones of more than 10 million ha each in West and Central Africa, for a total of 492 million hectares (Figure 8.6, color section).

Regional (or level II) characterization of inland valley agroecosystems were planned within the "windows" of macro (level I) zones. For rapid characterization and mapping at a regional scale of such a large area, which is spread across a subcontinent, it was only feasible using high-resolution satellite images (Thenkabail and Nolte, 1995a; Thenkabail and Nolte, 1996). The location of the acquired Landsat TM and SPOT HRV satellite images for regional (level II) characterization study in relation to the IITA level I (macroscale) map are displayed in Figure 8.6 in color section.

Four international research centers (IITA, WARDA, CIRAD, Winand Staring Centre and Wageningen Agricultural University) and several national research systems (from Republic of Benin, Burkina Faso, Côte d'Ivoire, Ghana, Mali and Nigeria) acknowledged this holistic research approach to characterization of IVs at a workshop at WARDA headquarters, Bouaké, Côte d'Ivoire, June 1993, since it fit well with the ideas developed by these institutes.

RATIONALE AND JUSTIFICATION

Characterization of the agroecosystems in which IVs occur is available at subcontinental/macrolevel of West and Central Africa (Hekstra et al., 1983; Andriesse and Fresco, 1991; Izac et al., 1991; Windmeijer and Andriesse, 1993). Country/mesolevel studies of inland valley agroecosystems were published for the Republic of Benin by Kilian (1972), for Senegal by Bertrand (1973), for Burkina Faso and Mali by Albergel et al. (1993), and for Côte d'Ivoire by Becker and Diallo (1992). Turner (1985, 1977) gives mesolevel details upon biophysical data of fadamas in Central and Northern Nigeria, which were completed by the studies of Kolawole (1991) on fadama economics and management. Many studies upon dambos and their cultivation pattern have been done in South and Southeastern Africa. For example, Rattray et al. (1953) described vlei cultivation in Rhodesia/Zimbabwe. Other regional studies in this part of sub-Saharan Africa are reviewed by Ingram (1991).

Given the fact that the characteristics of IVs are known to vary dramatically within and across agroecosystems, it is inadequate to characterize only a few IVs as most conventional studies often do. Lack of representativeness of a few study sites in the context of a regional agroecological zone leads to limited extrapolation or regionalization of the results of key sites to other areas within the same agroecological zone.

Lack of an appropriate approach to characterization constitutes a major constraint to research activities in developing technologies that are able to sustain the resource base of large regions and are adoptable by diverse groups of farmers with highly differing socioeconomic and ethnic backgrounds. Therefore, IITA attempted to systematically characterize agroecosystems of IVs in West and Central Africa. This three-tier methodology from macro- to microlevel is intended to lead to the development of technologies at benchmark sites.

OBJECTIVES AND APPROACH

Level-II (regional) characterization work involves establishing detailed characteristics of IV agroecosystems in sample areas of level I. The specific objectives envisaged in level-II characterization fall within the overall objectives of Izac et al. (1991) and were:

1. inventory the area of inland valley systems and their uplands;
2. map the spatial distribution of IVs, and study their spatial variability;
3. study the existing land-use pattern of IVs and uplands, and establish their interactions;
4. explore the potential of IVs for dry-season cropping;
5. determine the existing crop types and cropping pattern in IVs, and on uplands;
6. map the major road systems and significant settlements;
7. establish the cultivation pattern with respect to distance from road network and settlements;
8. study the watershed, and establish morphometric characteristics of IV watersheds;

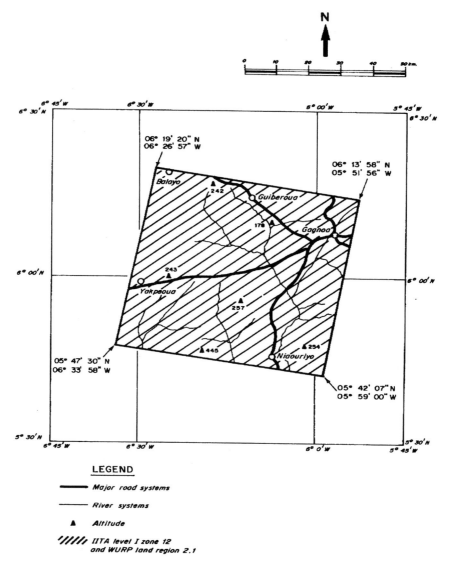

Figure 8.2 Spatial distribution of level 1 agroecological and soil zones in SPOT K:J of 47/338. Of the total area of).49 Mha, the entire 100% are falls in AEZ 16.

9. describe morphological IV characteristics such as shape, size, and slope;
10. make it possible to compare and contrast IVs within and across agroecological and socioeconomic zones in West and Central Africa;
11. rationalize the selection of representative benchmark site/s or benchmark IV watershed/s for the main phase research (i.e., technology development) activity; and
12. reveal the socioeconomic conditions of IV use. (Note that this is not part of the ongoing study).

IITA adopted a specific approach to achieve the above specific level-II objectives through integration of high-resolution satellite data with Global Positioning System (GPS) data, and ground-truth data in a Geographic Information Systems (GIS) framework. A comprehensive methodology

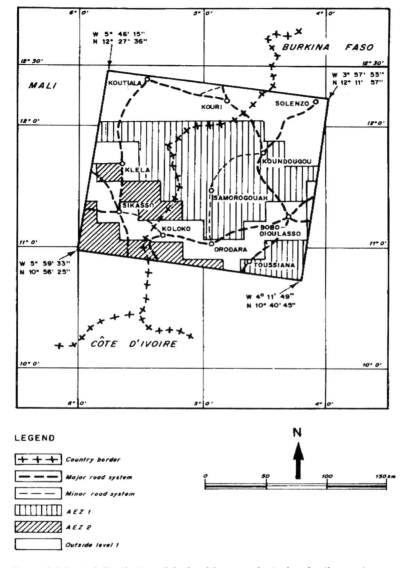

Figure 8.3 Spatial distribution of the level I agroecological and soil zones in Landstat 197/52. Of the total area of about 3.14 Mha., 45% is in AEZ 1 and 12% in AEZ 2. As for the rest of the area, 43% is outside the 18 AEZs mentioned in Figure 8.1.

for rapid characterization and mapping of inland valley systems at regional level using satellite imagery was developed and reported by Thenkabail and Nolte (1996, 1995a).

In this chapter two Landsat TM images (path 192 and row 54 and path 197 and row 52, [(see Figure 8.1and 8.2; and Table 8.1 for details and spatial location)], and one SPOT HRV image (Figure 8.3, 8.6 in the color section and Table 8.1)] were was used for the study. These images were selected, based on a set of criteria that included availability of cloud-free (0% cloud cover) satellite scenes, location of images in different agroecological and socioeconomic zones, most recently available dates of satellite overpasses (images were of early 1990's-see Table 8.1), accessibility of the region for ground-truthing, and coverage during wet and dry seasons. The images covered four agroecological and soil zones (AEZ) of level I—AEZ 1, 2, 7, and 16.

Inland valleys, their fringes, and uplands distributed in each of the above AEZs were be reported along with their land-uses, cultivation intensities , and their spatial distribution.

The level II work offers :

a. the capability to zoom in on IV agroecosystems from macro- to microlevels ("top-down" approach) leading to the rationalization of benchmark site selection for technology development research activities; and

b. the capability to extrapolate research results to larger regions or the regionalization of research results ("bottom-up" approach) for the transfer of technology.

DEFINITION USED FOR MAPPING INLAND VALLEYS

It is obvious from the various accounts in the literature that there is no single, widely accepted definition of IVs. They (IVs) are one of the many forms of wetlands that are characterized in their bottom by hydromorphic soils. Beyond this, there are significant variations in the definitions and characteristics of IVs even within the same region as perceived by different people (e.g., Savvides, 1981; Raunet, 1982; Hekstra et al., 1983; Acres et al., 1985; Mäckel, 1985; Turner, 1985; Andriesse, 1986; Oosterbaan et al., 1987; Andriesse and Fresco, 1991; Izac et al., 1991; Mokadem, 1992). For example, Mäckel (1985) defines a uniform zonation of dambos in Southeast Africa depending on vegetation, soil type, moisture content, and morphodynamics of the dambo.

Lack of a consistent definition of IVs has been a constraint in comparing or synthesizing different studies. To most researchers, characterization of IVs meant valley bottom characterization. Others included fringes also in their studies. Andriesse and Fresco (1991) proposed a physiohydrographic model of an inland valley.

The working definition adopted in this study will be as follows: Inland valleys (IVs) comprise valley bottoms and valley fringes (Figure 8.5). Valley bottoms are characterized by hydromorphic soils that constitute a relatively flat surface with or without a central stream. Valley fringes refer to areas along the slopes of the valley; rainfall either runs off above the surface of these areas or interflows horizontally on impervious subsurface layers toward the valley bottom and the central stream. Valley fringes, typically, have two distinct characteristic zones (Figure 8.5):

1. the lower part of the valley fringe immediately adjoining the bottoms that may have a high likelihood of a seasonal hydromorphic zone with significant potential for dry-season cropping; and

2. the upper part of the valley fringe with steeper slopes, in zones with less than 1400 mm rainfall (Guinea savanna) characterized predominantly by impervious layers (ironstones or carapaces) from which rainwater quickly runs off to the valley bottom). Soils in these upper portions of valley fringes dry out rapidly once the rains have ceased, and therefore, have no potential for dry-season cropping.

METHODOLOGY

An overview of the methodology has been shown in Figure 8.4 and described in detail in Thenkabail and Nolte (1995) and Thenkabail and Nolte (1996). The methodology permits a rapid characterization of large areas on a regional scale. The methodology includes the description of techniques for:

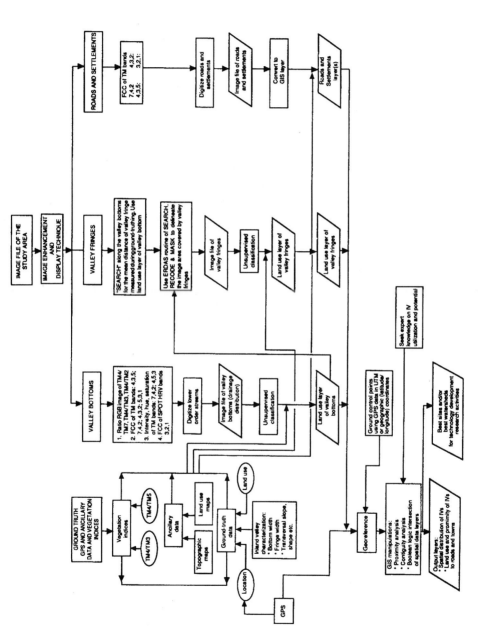

Figure 8.4 Methodology for mapping and characterizing Inland Valley (valley bottoms plus valley fringes) agroecosystems using Remote Sensing, Global Positioning System (GPS), and Ground-Truth Data.

1. valley bottom mapping;
2. valley fringe mapping;
3. mapping roads and settlements;
4. establishing land-use land-cover characteristics across the IV toposequence (uplands, valley fringes, and valley bottoms); and
5. establishing other IV characteristics such as IV stream densities and stream frequencies.

Figures 8.7 to 8.11 in the color section demonstrate the application of the methodology (Figure 8.4) for certain subareas of the study. Using the methodology, valley bottoms were highlighted (Figure 8.7, see color section) and delineated (Figure 8.8, see color section) for a subscene in Landsat TM path and row 192/54 that is part of agroecological and soil zone (AEZ) 7. The fringes adjoining the bottoms were mapped as illustrated in Figure 8.9 in the color section. A similar procedure was used to delineate valley bottoms from a SPOT subscene (Figure 10, see color section) which is in AEZ 16. Figure 8.11 in the color section (in AEZ 2) illustrates the large widths of the valley bottoms that are seasonally flooded and are characteristically dissimilar to valley bottoms in other AEZs. Sixteen land-use classes (Table 8.2) were mapped consistently across each study area based on the percentage distribution of 10 different land-cover types (Table 8.3).

Table 8.1 Parameters Describing the Level I Agroecological and Soil Zones[a]

Level I AEZ[b]	Agroecological zone According to IITA's Definition	LGP[c] (days)	Major FAO Soil Grouping[d]	Area[e] (million ha)
1	**Northern Guinea savanna**	**151–180**	**Luvisols**	**25.2**
2	**Southern Guinea savanna**	**181–210**	**Luvisols**	**18.4**
3	Southern Guinea savanna	181–210	Acrisols	12.4
4	Southern Guinea savanna	181–210	Ferralsols	11.9
5	Southern Guinea savanna	181–210	Lithosols	10.7
6	Derived savanna	211–270	Ferralsols	47.2
7	**Derived savanna**	**211–270**	**Luvisols**	**24.9**
8	Derived savanna	211–270	Nitosols	14.2
9	Derived savanna	211–270	Arenosols	14.0
10	Derived savanna	211–270	Acrisols	11.7
11	Derived savanna	211–270	Lithosols	10.8
12	Humid forest	> 270	Ferralsols	150.1
13	Humid forest	> 270	Nitosols	27.2
14	Humid forest	> 270	Gleysols	19.2
15	Humid forest	> 270	Arenosols	18.9
16	**Humid forest**	**> 270**	**Acrisols**	**18.0**
17	Mid-altitude savanna[b]		Ferralsols	45.4
18	Mid-altitude savanna[g]		Nitosols	12.3

[a]AEZs in bold have been investigated in this study at next level (level II) using high resolution satellite imagery, and the results are reported here
[b]AEZ: level I agroecological and soil zones.
[c]LGP: length of growing period.
[d]Names refer to the FAO soil classification scheme of 1974 (FAO/UNESCO 1974).
[e]The area figures are for West and Central Africa.
[f]Area distribution of LGP in AEZ 17 is: 151–180 days 11%, 181–210 days 9%, 211–270 days 59%, > 270 days 21%.
[g]Area distribution of LGP in AEZ 18 is: 151–180 days 2%, 181–210 days 5%, 211–270 days 53%, > 270 days 40%.

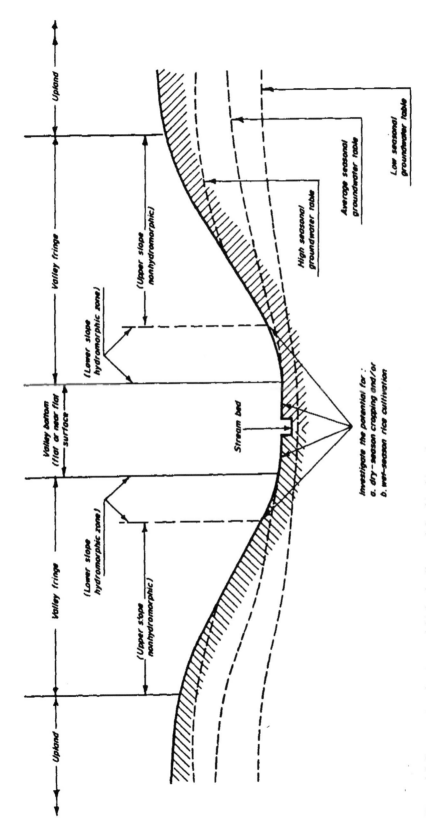

Figure 8.5 Cross-section showing model inland valley as defined in this study.

Table 8.2 Land-Use classes Mapped in Level-II Characterization of IV Agroecosystems of West and Central Africa

Code	Land-Use Class Description	Designated Color Derived Vegetation Indices
	Upland	
1	significant farmlands	gray
2	scattered farmlands	seafoam
3	insignificant farmlands	violet
4	wetland/marshland	mocha
5	dense forest	rose
6	very dense forest	red-orange
	Valley Fringe	
7	significant farmlands	white
8	scattered farmlands	pine-green
9	insignificant farmlands	red
	Valley Bottom	
10	significant farmlands	cyan
11	scattered farmlands	yellow
12	insignificant farmlands	magenta
	Others	
13	water	blue
14	built-up area/settlements	tan
15	roads	navy
16	barren land or desert land	sand

Table 8.3 Land-Cover Types Identified in Level-II Characterization of IV Agroecosystems of West and Central Africa

Code	Land-Cover Type Description	Code	Land-Cover Type Description
1	water	6	barren farms
2	tree	7	barren lands
3	shrub	8	built-up area/settlement
4	grass	9	roads
5	cultivated farms	10	others

STUDY AREAS AND GROUND-TRUTHING

In this chapter results of the three study areas (each satellite image comprising a study area) have been discussed (Table 8.4 and 8.1; Figures 8.1, 8.2, and 8.3; Figure 8.6, see color section). These study areas cover four distinct agroecological and soil zones (AEZ) in:

1. AEZ 2 (49% of the total area) and AEZ 7 (46%) by Landsat path 192 and row 54 (having an total area of 3.12 million hectares) (Figures 8.1, 8.6, see color section, Table 8.1 and Table 8.4);
2. AEZ 16 (100% of the total area) by SPOT HRV K 47 and J 338 (having an total area of 0.39 million hectares); (Figures 8.2, 8.6, see color section, Table 8.1 and Table 8.4) and
3. AEZ 1 (45% of the total area) and AEZ 2 (12%) by Landsat path 197 and row 52 (having an total area of 3.14 million hectares) (Figure 8.3, 8.6, see color section, Table 8.1 and Table 8.4).

Table 8.4 Attributes of Landsat TM and SPOT HRV Data Acquired for the Level II Characterization of Inland Valley Agroecosystems in West and Central Africa [a]

Landsat TM Path	Row		Center Coordinates		Date of
SPOT HRV					
K	J	Region / Country	longitude	latitude	Overpass
192	54	Save / Republic of Benin	02° 06′ E	08° 41′ N	12/29/91
47	338	Gagnoa / Côte d'Ivoire	06° 12′ W	06° 00′ N	12/27/90
197	52	Sikasso / Mali	04° 59′ W	11° 34′ N	09/27/91

[a](See their location in Figure 8.1)

Ground-truthing was conducted to correspond with satellite overpass dates during 1993 and 1994 (Table 8.4). The location of each ground-truth site was determined using a global positioning system (GPS) Garmin 100-SRVY. Locations noted were geographic co-ordinates (latitude/longitude) in degree, minutes and seconds, and universal transverse mercator (UTM) coordinates (x,y) in meters. The accuracy of these GPS readings was within ± 30 m. The GPS was also used to collect ground-control points to georeference the satellite image.

Land-use measurements were made in a 90 m by 90 m plot of the valley bottoms, valley fringes, and uplands. Its location was determined from the center with the GPS. The leaf area index of trees, shrubs, and grasses was measured in the same plot. Each land-use class has varying degrees of land cover types (Table 8.3). A total of 10 land-cover types was recorded at each ground-truth site: water, trees, shrubs, grasses, cultivated farms, barren farms, barren lands, built-up areas or settlements, roads, and others. Different combinations of these landcover types constitute specific land-use categories (Table 8.2).

The measurements (in meters) of valley-bottom and valley-fringe width were taken at a transect covering the 90 m by 90 m plot for land-use measurements. Other characteristics recorded at each inland valley sample site included valley bottom width (in meters), valley fringe width, transversal slope (degree), stream order (number), and qualitative observations of the soil moisture status as well as the level of water management system prevailing in valley-bottom fields.

RESULTS AND DISCUSSIONS

The results of each of the three study areas have been discussed seperately in the subsections below. Data from Landsat and SPOT satellites for the study areas described below were integrated with GPS location data and ground-truth data in a GIS framework. The spatial data layers were georeferenced to Universal Transverse Mercator (UTM) coordinates obtained during ground-truthing using a GPS. Digital georeferenced databases generated include spatial distribution of inland valleys and land-use classes of inland valleys and uplands.

Save (Republic of Benin) study area

The study area [Figures 8.1, 8.6 (see color section), Table 8.1, and 8.4] encompassed two level I (macro or subcontinental) agroecological and soil zones: 1.AEZ 2 (southern Guinea savanna with Luvisols) and AEZ 7 (derived savanna with Luvisols) [see Table 8.1, Figures 8.1, and 8.6 (see color section)]. AEZ 2 is representative of 18.4 million ha and AEZ 7 of 24.9 million ha in West and Central Africa (see spatial distribution of these zones in Figure 8.6 in the color section and Table 8.1). Both these agroecological and soil zones have Luvisols as major soil grouping. The zones differ in their length of growing period, with AEZ 2 having 181–210 days and AEZ 7 having 211–270 days. The study area is part of a region that gets rainfall varying between 975 mm and 1336 mm per year. Of the total study area 49% falls into AEZ 2 and with 46% into AEZ 7, while the remaining 5% is located outside the level I zones.

A total of 9.0% (281,500 ha) of the entire study area (3.12 million ha) is occupied by valley bottoms (Figure 8.12, see color section and Table 8.5). The corresponding numbers for the sub-zones were 7.9% in AEZ 2 and 10.2% in AEZ 7. Valley fringes accounted for 18.3% in AEZ 2, 21.8% in AEZ 7, and 19.9% in the entire study area, while uplands represented 73.3% in AEZ 2, 67.1% in AEZ 7, and 70.3% in entire study area. These results indicated a greater percentage of inland valleys in the wetter zone (AEZ 7) compared to the drier zone (AEZ 2).

The cultivation intensity differed only marginally between the AEZs (Table 8.5). A total of 7.9% of the valley bottoms in the entire study area were cultivated with 7.9% in AEZ 2 and 7.7% in AEZ 7. A total of 15.9% of the valley fringes was cultivated in the entire study area with 15.1% in AEZ 2 and 16.6% in AEZ 7. Compared to the valley fringes, less of the upland area was used as farmland, 13.2% across the entire study area, 11.2% in AEZ 2, and 14.3% in AEZ 7. These results indicated a low percentage of cultivation across the toposequence. This was attributed to: (a) a low density of the road network and (b) a low number of settlements, hence low population densities. In addition, socioeconomical issues such as the traditional preference of farmers in cultivating uplands, debilitating diseases in lowlands, lack of appropriate technologies (e.g., low-cost water management techniques), lack of knowledge about lowland agriculture, lack of market access, and lack of governmental support for farmers imply less than optimal utilization of inland valleys for agriculture.

TM data facilitated mapping the shape, size, spatial distribution, density, and frequency of inland valley bottoms occurring in the study area. The spatial distribution of these characteristics is illustrated in Figure 8.12 in the color section. They were also found to be very useful in detecting settlements (usually of sizes of >15 ha) and the motorable road network. This was especially useful in mapping the recent spread of settlements and the road systems in the absence of such information on base maps, such as topographic maps (which were mostly available from the early 1960s). Furthermore, the cultivation pattern of inland valleys with respect to their distance from settlements and the road network could be studied.

There was a high spatial correlation in the cultivation pattern of valley bottoms and valley fringes compared with the cultivation pattern of uplands. Significantly cultivated areas are mainly concentrated along major road networks and near settlements. The degree of cultivation decreased significantly as the distance from settlements or the road network increased. Stream drainage densities (km/km^2) were denser in the wetter zone (AEZ 7: 1.07 to 1.20 km/km^2) compared to the drier zone (AEZ 2: 0.9 to 1.0 km/km^2). The same was true for inland valley stream frequencies (number of streams per km^2): 0.95 to 1.10 streams/km^2 in AEZ 7; and 0.74 to 0.79 streams/km^2 in AEZ 2.

About 90% of the inland valleys were classified as U-shaped. Only about 17% of those surveyed during ground-truthing had some improved water management schemes, such as field leveling and bunding. The valley bottom widths of first- to third-order valleys in AEZ 2 and AEZ 7 varied between 40 m and 81 m and hence provided considerable valley-bottom areas (7.7% to 9.0% of the geographic area). The valley bottoms are flat to near-flat. Most valley bottoms are characterized by grassy vegetation that could be easily cleared for cultivation.

Land-use characteristics were mapped separately for valley bottoms, valley fringes, and up-

Table 8.5. Distribution of Valley Bottoms, Fringes, and Uplands, and Their Cultivation Pattern in the Save (Republic of Benin) Study Area[a]

	Subarea	Level I Agroecological and Soil Zones	% of Entire Study Area (% of Full Scene)	Valley-Bottom Area		Valley-Fringe Area		Upland Area	
				as a % of Total Geographic Area (%)	Cultivated as a % of Total Valley-Bottom Area (%)	as a % of Total Geographic Area (%)	Cultimated as a % of Total Valley-Fringe Area (%)	as a % of Total Georgraphic Area (%)	Cultivated as a % of Total Upland Area (%)
A.	Parakou and Tchaourou region of Rep. of Benin	AEZ 2	49	7.9	7.9	18.3	15.1	73.3	11.2
B.	Save/Bante/Gonka/Bassila region	AEZ 7	46	10.2	7.7	21.8	16.6	67.1	14.3
C.	Entire study area	land region 2.1	100	9.0	7.9	19.9	15.9	70.3	13.2

Notes:

[a]AEZ 2 represents the southern Guinea savanna with Luvisols and AEZ 7 the Derived savanna with Luvisols. AEZ 2 occupies 49% and AEZ 7 46% of the entire study area. The remaining 5% is located outside the level I zones.

Table 8.6 Distribution of Land-Use Classes in the Different Agroecological and Soil Zones of the Save (Republic of Benin) Study Area[a]

	Land-Use Class	AEZ 2		AEZ 7		Entire Study Area	
		Area (ha)	% of Total AEZ 2	Area (ha)	% of Total AEZ 7	Area (ha)	% of Total Area
	Uplands						
1	significant farmlands	57,853	3.8	148,609	10.3	221,399	7.1
2	scattered farmlands	547,694	35.7	492,119	34.2	1,099,828	35.2
3	savanna vegetation[b]	233,120	15.2	155,091	10.8	407,026	13.0
4	wetlands/marshland	129,808	8.5	78,665	5.5	218,185	7.0
5	dense forest	126,096	8.2	63,876	4.4	194,141	6.2
6	very dense forest	30,498	2.0	28,445	2.0	60,052	1.9
	Valley fringes						
7	significant farmlands	12,550	0.8	32,452	2.3	47,861	1.5
8	scattered farmlands	234,627	15.3	250,003	17.4	507,711	16.2
9	insignificant farmlands[c]	33,846	2.2	31,807	2.2	67,597	2.2
	Valley bottom						
10	significant farmlands	2,012	0.1	4,943	0.3	7,537	0.2
11	scattered farmlands	78,174	5.1	82,934	5.8	169,027	5.4
12	insignificant farmlands[d]	41,813	2.7	59,160	4.1	103,895	3.3
	Others						
13	water	402	0.0	629	0.0	1,223	0.0
14	settlements/built-up area	668	0.0	1,279	0.1	4,967	0.2
15	roads	1,856	0.1	1,858	0.1	3,929	0.1
16	barren/desert area	3,938	0.3	8,324	0.6	13,999	0.5
	Total	1,534,955	100.00	1,440,193	100.00	3,128,377	100.00

[a]Level I agroecological and soil zones (see Figure 8.1 and Table 8.1). AEZ 2 represents the southern Guinea savanna with Luvisols and AEZ 7 the Derived savanna with Luvisols. AEZ 2 occupies 49% and [b]AEZ 7 46% of the entire study area. The remaining 5% is located outside the level I zones.
[c]Class 3 occurs only in Guinea savanna zones.
[d]Spectral characteristic of vegetation in class 9 is similar to that of classes 5, 6, and 12; the difference is mainly in the toposequence position,
[e]Mainly riparian vegetation; spectral characteristics of vegetation similar to classes 5, 6, and 9; the difference is mainly in the toposequence position,

lands using the CLUSTR unsupervised classification algorithm of ERDAS (ERDAS, 1992). The areas of each of the 16 classes (Figure 8.15, see color section) in the two agroecological and soil zones and in the entire study area are given in Table 8.6. These land use classes are designed on the lines of Anderson et al. (1976). Each land-use class contains a varying degree of land-cover types (Table 8.3).

Gagnoa (Cote d'Ivoire) Study Area

In this study SPOT–1 HRV data for location K:47, J:338 (of the SPOT grid reference system; see precise coordinates in Figure 8.4) with 27 December 1990 as the date of overpass covering an area of 393,112 ha in the Gagnoa region of southwestern Côte d'Ivoire (Figure 8.4).

Table 8.7. Distribution of Valley Bottoms, Valley Fringes, and Uplands and Their Cultivation Status in the Gagnoa (Côte d'Ivoire) Study Area[a]

Entire Study Area	Valley Bottom Area		Valley Fringe Area		Upland Area	
	as a % of Total Geographic Area (%)	Cultivated as a % of Total Valley-Bottom Area (%)	as a % of Total Geographic Area (%)	Cultivated as a % of Total Valley-Fringe Area (%)	as a % of Total Geographic Area (%)	Cultivated as a % of Total Upland Area (%)
100	18.0	20.6	40.3	16.9	40.1	15.0

[a]The study area falls entirely into agroecological zone 16 of the level I map (Figure 8.1 and Table 8.1).

The entire study area [Figure 8.2, and 8.6 (see color section) Table 8.1, and Table 8.4] falls within a single level I (macro) agroecological and soil zone 16 (AEZ 16): the humid forest zone with Acrisols as the major soil grouping according to the FAO soil map of the World (FAO/UNESCO, 1977) (see Figure 8.2 and Table 8.1). As a result, the characteristics reported for the entire study area will also be true for the macroscale zones AEZ 16 (Figure 8.2).

The study area comprises 18% valley bottoms (Figure 8.13, see color section), 40.3% valley fringes, 40.1% uplands, and 1.6% others (e.g., roads, settlements) (Table 8.7). The area covered by valley bottoms (18%) was relatively high as a result of large bottom widths with mean values of 89 (first-order valleys), 164 m (second-order), 173 m (third-order), and 244 m (fourth-order). These measurements showed a high variability across the stream orders. First-order valleys had significantly smaller valley bottoms than higher-order valleys. The inland valley stream density in the study area was determined as 0.97 km/km^2 (medium as per Hekstra et al., 1983) and the stream frequencies were determined to be 0.75 streams per km^2 (coarse as per Hekstra et al., 1983). The mean valley fringe widths were 300 m (first-order valleys), 303 m (second-order), 355 m (third-order), and 436 m (fourth-order).

The 16 land-use classes (Figure 8.16, see color section)– six for uplands, three each for valley fringes and valley bottoms, and four others—were mapped for the entire study area (Table 8.8). The total area covered by humid forest vegetation with insignificant farmlands (land-use classes 5, 6, 9, and 12) was 58.3%, whereas the area with humid forest-cropland mosaic (land-use classes 2, 8, and 11) was 23.0% of the total geographic area. The intensity of cultivation was significantly higher for valley bottoms (20.6%) compared with valley fringes (16.9%), and uplands (15%) (Table 8.7). This was mainly a result of the use of valley bottoms for rice cultivation, as revealed by ground-truthing. In the entire study area of the image, the most significant areas of farmlands were in the lower left corner (southwestern portion) of the image near the settlement of Yakpéoua and around predominant settlements, such as Gagnoa and Guibéroua (Figure 8.4). In significant portions of the image there was a strong relationship in the cultivation of uplands and inland valleys.

As expected in a low population density area that belongs to the humid forest zone (Figure 8.6 in color section), the forest vegetation (classes 5, 6, 9, and 12) dominates the image. In the classification system used in this study, the characteristics of areas with insignificant farmlands at valley fringes (class 9) and in valley bottoms (class 12) are similar to those areas with dense and very dense forest on uplands (classes 5 and 6). The only difference is in the toposequential position of the area under consideration. The vegetation of these classes mostly consists of dense and vigorous evergreen trees, and dense and vigorous shrubs. Grasses were less common and farmlands were insignificant in these classes.

Table 8.8 Land-Use Distribution in the Gagnoa (Côte d'Ivoire) Study Area[a]

No.	Land-Use Category	Full Study Area Area (ha)	% of Total Study Area
	Uplands	157,601	40.1
1	significant farmlands	22,589	5.8
2	scattered farmlands	31,992	8.1
3	savanna vegetation[b]	0	0
4	wetlands/marshland	7,024	1.8
5	dense vegetation	54,619	13.9
6	very dense vegetation	41,377	10.5
	Valley fringes	158,606	40.3
7	significant farmlands	26,299	6.7
8	scattered farmlands	39,376	10.0
9	insignificant farmlands[c]	92,931	23.6
	Valley bottom	70,638	18.0
10	significant farmlands	11,490	2.9
11	scattered farmlands	19,058	4.9
12	insignificant farmlands[d]	40,090	10.2
	Others	6,268	1.6
13	water	358	0.1
14	built-up area/settlements	2,703	0.7
15	roads	2,194	0.5
16	barren land or desert lands	1,013	0.3

Notes:
[a]The study area falls entirely into agroecological zone 16 of the level I map (Figure 8.1 and Table 8.1)
[b]Class 3 occurs only in Guinea savanna zones.
[c]Spectral characteristic of vegetation in class 9 is similar to that of classes 5, 6, and 12; the difference is mainly in the toposequence position.
[d]Mainly riparian vegetation; spectral characteristics of vegetation similar to classes 5, 6 and 9; the difference is mainly in the toposequence position.

Areas on uplands, at fringes, and in valley bottoms with forest and savanna vegetation and scattered farmlands (classes 2, 8, and 11) have > 10% and < 30% of cultivation and can be termed as forest-cropland mosaic. This mosaic occurs especially in the middle and western portions of the image, covers about two-thirds of the overall image, and is typically to be found around settlements and major road networks (areas around Gagnoa and Guibéroua in Figure 8.4).

The forest-cropland mosaic land-use classes (2, 8, and 11) contain a significantly higher percentage of farmlands (land-cover types 5 and 6) and a lower percentage of trees and shrubs (land-cover types 2 and 3), but they still have a significant amount of forest vegetation. The overall area under forest-cropland mosaic represents 23.0% of the total geographic area.

Ninety-two percent (92%) of the inland valleys (valley bottoms plus valley fringes) were U-shaped. Ninety percent of these valleys were considered suitable for cultivation, which means that the bottom width was reasonably large enough (> 5–10 m wide and more or less flat). Low-level water management systems, such as leveling and bunding of fields were found in 25% of all the valleys. The characteristics of valley bottoms, such as their widths, spatial distribution (density, frequency, and location), and land-use characteristics, were mapped for the entire study area as displayed in Figure 8.16, see color section.

The characteristics of inland valleys in the region clearly indicate a vast potential for agricultural development (Table 8.7): over 80% of the valley bottoms and valley fringes are currently unexploited, they offer wide bottom widths, adequate water resource, and are relatively easy to clear compared to uplands due to the presence of significantly fewer trees.

All major and minor roads discernible on SPOT data were mapped, providing a road network density of 0.16 km/km^2 which is considered as a good road infrastructure by Manyong et al (in preparation). Ninety-three settlements of varying sizes were also mapped based on SPOT data, constituting a significant portion of the overall settlements in the study area. The relationship between intensity of cultivation and the distance of farmlands from roads and settlements was established. The intensity of cultivation at all the components of the toposequence (valley bottoms, valley fringes, and uplands) was slightly lower at distance zones beyond 4 km compared with distances within 4 km. This was due to the presence of a large number of well connected small road systems, equally well distributed small settlements, and therefore good access to markets. The spatial distribution of these characteristics are illustrated in Thenkabail and Nolte (in press, 1995c).

Sikasso (Mali), Bobo-Dioulasso (Burkina Faso) Study Area

In this study area Landsat–5 Thematic Mapper (TM) Path: 197, Row: 52, with a total area of 3.13 million hectares, (see the exact coordinates in Figure 8.3) was used. The total study area (3.13 million hectares) [Figure 8.3, 8.6 (see color section) Table 8.4] comprised 8.6% valley bottoms, 20.4% valley fringes, and 70.2% uplands (Table 8.9). Water bodies, roads, and settlements comprised the other 0.8% area (Table 8.10). The drainage density of 0.4 km per km^2, and stream frequency of 0.61 number per km^2 obtained in the study area were classified as low (0.3 to 0.6 km/km^2), and coarse (0.5 to 1.0 number/km^2), respectively, by Hekstra, et al. (1983). In spite of the low and coarse drainage densities, and stream frequencies in the study area, the percentage area of inland valleys (valley bottoms plus valley fringes) were significant mainly as a result of large valley bottom and fringe widths of the inland valley streams (first- to fourth-order streams). The mean bottom widths for the first- to third-order stream were about 90 meters, increased dramatically for the fourth-order to about 400 meters. The mean valley fringe (hydromorphic plus nonhydromorphic) widths for the first three inland valley streams were about 200 meters, and for the fourth-order stream about 920 meters. Hence, even though the stream frequencies and stream densities were coarse and low, respectively, the large sizes of the valley bottoms and valley fringes led to their significant percentages.

Due to significant differences in the geographical areas studied (45% of entire study area for AEZ 1, 12% for AEZ 2; Table 8.9) a direct and realistic comparison of results across zones was not feasible. However, it may be noted that the valleys in AEZ 1 had greater bottom widths than valleys in AEZ 2, resulting in higher percentage area of valley bottoms (Figure 8.14, see color section) in AEZ 1 (9.1 %) compared to AEZ 2 (7.7 %). Due to the same reason, the percentage valley bottom area in land region 2.8 (9.1%) exceeded that of land region 3.3 (8.2%).

The study mapped 16 land use classes [Figure 8.17 (see color section) and Table 8.10] in the total study area of 3.13 million hectares and for the subzones [Figure 8.19 (see color section)]. The spatial distribution of these characteristics are illustrated in Thenkabail and Nolte (1995d).

The grassland dominant savannas dominate the study area. The overall savanna percentage areas were 65.5% for AEZ 1, 70.4% for AEZ 2, 67.1% for land region 3.3, 61.6% for land region 2.8, and 65.4% for the entire study area. The forest classes are predominantly trees along the river bank and were about 4% for all level I zones within the study area. This very low percentage of

Table 8.9. Distribution of Valley Bottoms, Valley Fringes, and Uplands and Their Cultivation Status in the Sikasso (Mali), and Bobo-Dioulasso (Burkina Faso) Study Area

Study Area	% of Entire Study Area	Valley Bottom Area		Valley Fringe Area		Upland Area	
		as a % of Total Geographic Area Scene)	Cultivated as a % of Total Valley-Bottom Area (%)	as a % of Total Geographic Area (%)	Cultimated as a % of Total Valley-Fringe Area (%)	as a % of Total Georgraphic Area (%)	Cultivated as a % of Total Upland Area (%)
AEZ 1[b]	45	9.1	15.9	17.8	16.0	72.2	20.5
AEZ 2	12	7.7	24.0	25.5	21.8	66.5	24.1
Entire study area	100	8.6	18.4	20.4	19.2	70.2	21.9

Notes:

[a]When valley bottoms + valley fringes + uplands are not equal to 100%, the rest of the area falls in water body, roads, and settlements or "round-off" errors.

[b]Level I agroecological and soil zones (see Figure 8.1).

Table 8.10. Land-Use Distribution in the Different Agroecological and Soil Zones (AEZ) Determined in the Sikasso (Mali), and Bobo-Dioulasso (Burkina Faso) Study Area

No.	Land-Use Class	AEZ 2		AEZ 7		Entire Study Area	
		Area (ha)	% of Total AEZ 2	Area (ha)	% of Total AEZ 7	Area (ha)	% of Total Area
	Uplands						
1	significant farmlands	105,468	7.4	32,124	8.3	280,990	9.0
2	scattered farmlands	367,370	25.9	141,550	36.4	901,130	28.7
3	savanna vegetation	406,472	28.7	63,360	16.3	727,987	23.2
4	wetlands/marshland	46,023	3.2	3,433	0.9	85,202	2.7
5	dense vegetation	51,818	3.7	12,078	3.1	79,955	2.6
6	very dense vegetation	10,797	0.8	1,172	0.3	18,175	0.6
	Valley fringes						
7	significant farmlands	31,561	2.2	17,789	4.6	104,879	3.3
8	scattered farmlands	95,177	6.7	52,260	13.4	282,455	9.0
9	insignificant farmlands	126,930	8.9	29,363	7.5	254,088	8.1
	Valley bottoms						
10	significant farmlands	7,717	0.5	4,863	1.2	21729	0.7
11	scattered farmlands	59,062	4.2	16,792	4.3	139,758	4.5
12	insignificant farmlands	62,045	4.4	8,726	2.2	107,519	3.4
	Others						
13	water	9,750	0.7	544	0.1	12,570	0.4
14	built-up area/settlements	299	0.0	82	0.0	5,849	0.2
15	roads	2,170	0.2	498	0.1	6,585	0.2
16	barren/desert area	35,643	2.5	4,600	1.2	106,985	3.4

forest cover was only to be expected in the study area as it falls in the northern guinea savanna and sudan savanna. The study areas in the sudan savanna comprised of maximum barren areas of 6% when compared to other level I zones studied.

The cultivation intensities were nearly the same across the toposequence with 18.4% for valley bottoms, 19.2% for valley fringes, and 21.9% for uplands (Table 8.9). The significant cultivation across the toposequence were mainly attributed to the market-driven conditions. In most cases cultivation intensities were about 3% higher for distance limits within 0–5 km from road network and settlements compared to those areas beyond 5 km.

The valley bottoms in the study area were characterized by flat or near-flat surfaces that have shallow flooding all through the rainy season. Rice cultivation (Figure 8.19, see color section) forms an important component of lowland rainy season cultivation in the entire study area, especially in the valleys surrounding Sikasso, Mali. These broad and flat or near-flat valley bottoms of the study area offer an excellent opportunity for innundated rice cultivation (swamp rice) during rainy season. However, the area of valley bottoms available for cultivation far exceeds their current utilization. A total of 269,006 hectares constitute valley bottoms (Figure 8.14, see color section) in the entire study area, of which only 18.4% (49,497 hectares) was cultivated. Of the cultivated IVs 42% (20,789 hectares) had rice cultivation (Figure 8.19, see color section).

One hundred percent of the inland valleys that were studied were 'U' shaped, 74% were fadamas (that is, inland valleys with potential for dry season cropping). At the time of ground-truthing 69% of the inland valleys were wet, 21% were moist, and 10% were dry. A mapping of biomass levels during summer will show: (a) upland and lowland contrasts in maximum as a result of significant moisture presence in lowlands relative to near dry uplands; (b) biomass variations within valley highlighting moisture availability in the valleys (see an illustration in Figure 8.21, see color section) when upland vegetation is dry. The mean transversal slopes were generally mild with about 1.5 degrees for the first- to third-order inland valley streams, and about 0.5 degrees for the fourth-order inland valley streams.

The study showed a strong relationship between upland cultivation and inland valley cultivation (Table 8.9) proving one of the hypotheses of this study.

SUMMARY

This study used high-resolution satellite images from Landsat TM and SPOT HRV in three separate study areas [Figure 8.1, 8.2, 8.3 and 8.6 (see color section), and Table 8.4) covering four agroecological and soil zones (AEZs; as shown in Figure 8.6 in the color section) of West and Central Africa to characterize and map inland valley agroecosystems. The characteristics studied include: inventorying the areas of inland valleys (valley bottoms plus valley fringes), determining the cultivation intensities of inland valleys and their uplands, and studying the land-use characteristics of inland valley bottoms and fringes, and their uplands.

The paper briefly outlined the holistic three-tier "top-down" approach developed by IITA to rapidly characterize and map inland valley agroecosystems of West and Central Africa at macro (level I), regional (level II), and micro (level III) scales. The focus of the paper was in applying the methodology for regional (level II) characterization and mapping of inland valleys in four major level I agroecological and soil zones (AEZ) using high-resolution satellite imagery from Landsat and SPOT systems (see Figures 8.7–8.11 in the color section for illustration of the methodology along with Figure 8.4). The three study areas were: (1). Save, Republic of Benin (Figure 8.1) covered by a Landsat TM image; (2). Gagnoa, Côte d'Ivoire (Figure 8.2) covered by a SPOT HRV; and (3). Sikasso, Mali/Bobo-Dioulasso, Burkina Faso (Figure 8.3) covered by a Landsat TM image. The three images cover four different agroecological and soil zones [AEZ 1, 2, 7, and 16; see Figures 8.1, 8.2, 8.3, and 8.6 (see color section), and Table 8.4] with a total area of about 6.6 million hectares.

The study made it possible to estimate inland valley (valley bottoms plus valley fringes) areas and their degree of cultivation (Table 8.5, 8.7, and 8.9). Inland valley bottoms (Figures 8.12, 8.13, and 8.14 in the color section) were highest in the humid forest zone (AEZ 16) with 18% (Figure 8.13 in the color section with Figure 2, and Table 7) of the total geographic area of study compared to 9.1% (Figure 8.14 in the color section with Figure 8.3, and Table 8.9) in northern Guinea savanna (AEZ 1), 7.7% (Figure 8.14 in the color section with Figure 8.3, Table 8.9) to 7.9% (Figure 8.12 in the color section with Figure 8.1, and Table 8.5) in southern Guinea savanna (AEZ 2), and 10.2% (Figure 8.12 in the color section with Figure 8.1, and Table 8.5) in the derived savanna (AEZ 7).

Currently, only 8% to 20% inland valley bottoms and 15% to 22% inland valley fringes are exploited for agriculture in AEZ 1, 2, 7, and 16 (Tables 8.5, 8.7, and 8.9). The vast percentage of unexploited inland valleys are rich in soil fertility, have potential for rice cultivation during main cropping season, and significant percentage of valleys have sufficient moisture to sustain upland crops during dry season.

Sixteen similar land-use classes were mapped cutting across the toposequence (uplands, valley fringes, and valley bottoms) for each study area (Tables 8.6, 8.8, and 8.10; and Figures 8.15, 8.16 and 8.17 in the color section). The toposequence oriented land-use mapping has been a unique feature of this study. It is indeed clear from this study that the inland valley bottoms that constituted

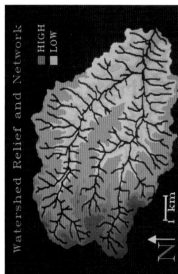

Figure 2.4. Generated elevation contours and channel network.

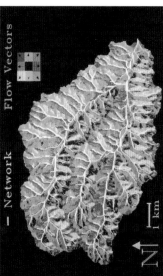

Figure 2.5. Generated flow vectors (or aspect) and channel network.

Figure 2.6. Watershed subdivision with corresponding channel network.

3-D Perspective North-West View

Figure 2.1. Three-dimensional, vertically enhanced DEM representation of Bills Creek watershed.

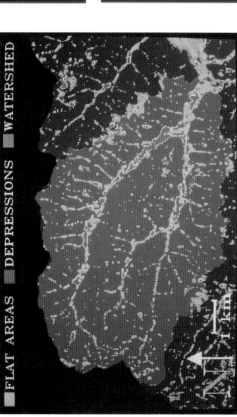

Figure 2.2. Location and extent of flat areas (green) and depressions (red) within the watershed area (blue).

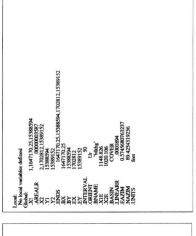

A Drainage-divide (cover_bas) digitized from topographic map.

Base from U.S. Defense Mapping Agency
1:250,000, 1976
Universal Transverse Mercator projection
Zone 15

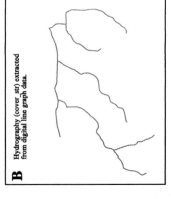

B Hydrography (cover_str) extracted from digital line graph data.

Base from U.S. Geological Survey digital data
1:100,000, 1984
Universal Transverse Mercator projection
Zone 15

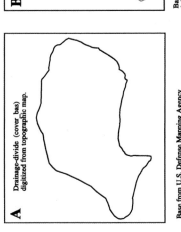

C Hypsography (cover_con) generated from digital elevation model data using ARC/INFO.

Base from U.S. Defense Mapping Agency
1:250,000, 1976
Universal Transverse Mercator projection
Zone 15

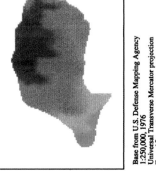

```
BASINNAME   =  BLKHG
TDA            56.955
NCDA            0.000
CDA            56.955
BL             11.441
BP             33.066
BS             52.319
BR            186.644
BA             89.425
BW              4.978
SF              2.298
ER              0.745
RB              1.804
CR              1.256
RR              5.645
MCL            16.819
TSL            73.896
MCS             8.674
MCSR            1.470
SD              1.297
CCM             5.771
RN            242.162
SR              0.166
FOS            28
BSO             3
DF              0.492
RSD             0.292
```

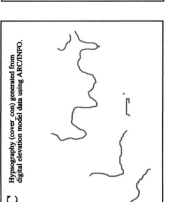

```
Local:
No local variables defined
Global:
X1        1,16471170.25,15388594
Y1        0000003587
AREALR    2,1702812,15389152
X2        15388594
Y2        15389152
IRDS      16471170.25,15388594,1702812,15389152
BX        15388594
BY        1702812
EX        15389152
EY        50
INTERVAL  '%bhg'
ORIENT
IBNAME    1148.826
X1E       1020.106
X2E       COVER
BASIN     00001894
LINEARR   0.574568076237
LFADM     89.4254319236
NAZIM     feet
UNITS
```

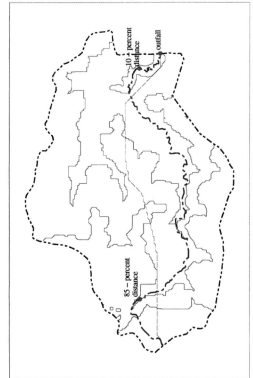

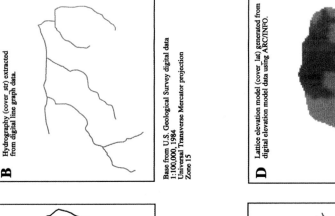

D Lattice elevation model (cover_lat) generated from digital elevation model data using ARC/INFO.

Base from U.S. Defense Mapping Agency
1:250,000, 1976
Universal Transverse Mercator projection
Zone 15

85 – percent distance
10 – percent distance
outfall

| 0 | 2 | 4 MILES |
| 0 | 2 | 4 KILOMETERS |

Contour interval = 50 ft

——— Basin contours
········· Basin–length measurement line
—·—·— Basin divide
— — — Main channel

Figure 4.5. Examples of graphical output from Basinsoft.

Figure 4.3 A–D. Examples of four source-data layers required by Basinsoft—(A) drainage-divide data layer; (B) hydrography data layer; (C) hypsography data layer; and (D) lattice elevation model data layer.

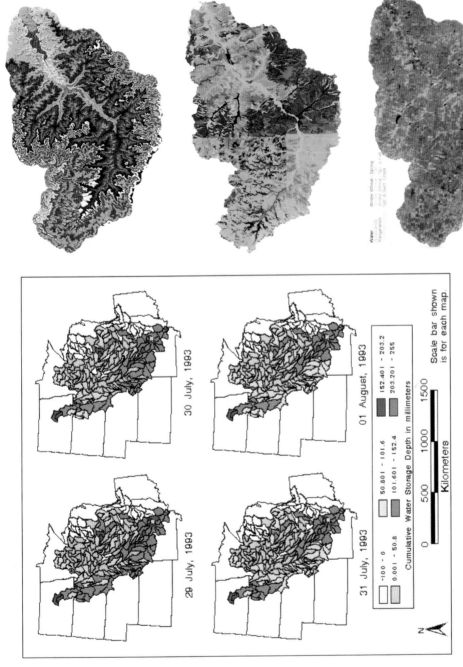

Figure 7.2. Digital elevation model of the Little Washita River Watershed overlain by TOPAZ-derived stream network. Each change in color on the DEM represents an elevation change of 7 m.

Figure 7.3. Soils coverage for the Little Washita River Watershed. Note that the soil mapping unit codes delineate the county boundaries.

Figure 7.4. April 1994 synoptic temporal land cover classification for the Little Washita River Watershed.

Figure 6.4. Cumulative water storage amounts within USGS HUC boundaries (spatially averaged values from gauging station zones) for the time period 07/29/93 - 08/01/93. Flooding peaked in the St. Louis area on 08/01/93. The city is shown by a small light-colored dot placed at the southern end of the basin.

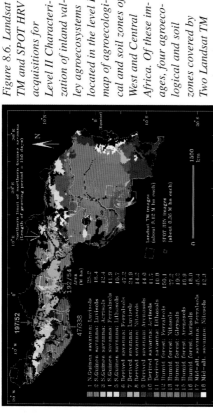

Figure 8.8. Inland valley bottoms delineated (for area shown in Figure 8.7) and displayed as a false color composite of TM bands: TM4 (red), TM3 (green), TM5 (blue). Color key: red (densest vegetation), shades of green and blue (very little and/or dry vegetation). The NDVI values were less than 0.10 for these areas.

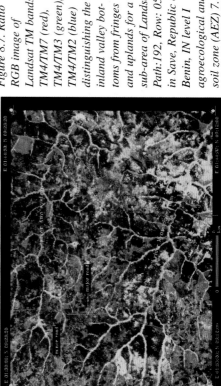

Figure 8.9. Inland valley fringes for the same sub-area as shown in Figures 8.7, and 8.8 displaying the false color image of TM4 (red), TM3 (green), TM5 (blue). Color key: red (densest vegetation), shades of green (lowest density and vigor of vegetation), and shades of blue and gray (intermediate vegetation).

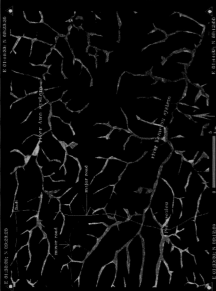

Figure 8.6. Landsat TM and SPOT HRV acquisitions for Level II Characterization of inland valley agroecosystems located in the level I map of agroecological and soil zones of West and Central Africa. Of these images, four agroecological and soil zones covered by Two Landsat TM (path and rows: 192/54; 197/52) and one SPOT HRV image (K and J: 47/338) have been used for illustration in this study.

Figure 8.7. Ratio RGB image of Landsat TM bands: TM4/TM7 (red), TM4/TM3 (green), TM4/TM2 (blue) distinguishing the inland valley bottoms from fringes and uplands for a sub-area of Landsat Path:192, Row: 054 in Save, Republic of Benin, IN level I agroecological and soil zone (AEZ) 7.

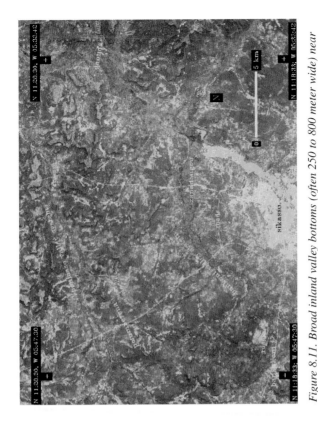

Figure 8.11. Broad inland valley bottoms (often 250 to 800 meter wide) near Sikasso, Mali. These valley bottoms are heavily flooded during rainy seasons and are either used for rice cultivation or have potential for rice and other wetland cropping. Characteristic valley bottoms and their uplands in AEZ 2.

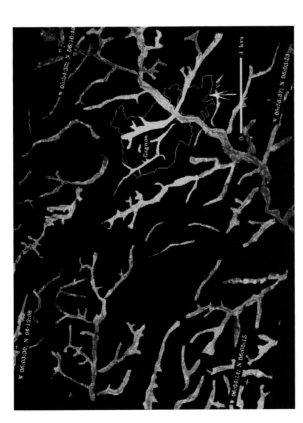

Figure 8.10. Delineated inland valley bottoms around Gagnoa, Cote d'Ivoire, displayed as a false color composite of SPOT HRV bands: band 3 (red), band 2 (green), and band 1 (blue). In AEZ 16.

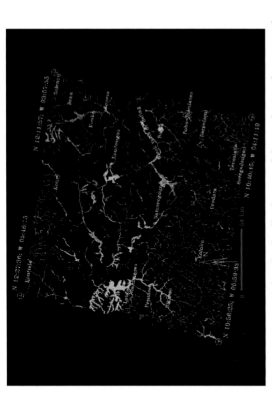

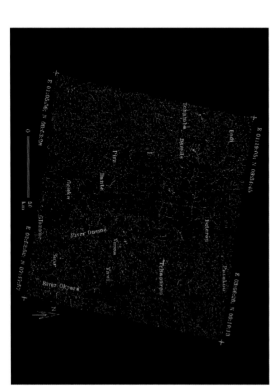

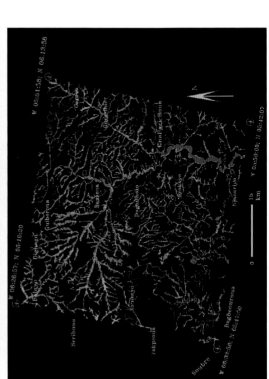

Figure 8.12. (top, left) Inland valley bottoms delineated using the espoused methodology for Landsat path:192 and row 54. There were 9 percent (281,500 ha.) valley bottoms (as shown in this plate) in total area of 3.14 Mha. The valley bottom area as a percent of total area was: 7.9 percent for AEZ 2 and 10.2 percent for AEZ 7. (see Figure 8.6 for spatial distribution of AEZ 2 and AEX 7 within this image; Color legend in Figure 8.18).

Figure 8.13. (left) Inland valley bottoms delineated using the espoused methodology for SPOT HRV K:47 and J:338. There were 18 percent (70,000 ha.) valley bottoms (as shown in this plate) in total area of 0.39 Mha. The valley bottom area as a percent of total area was: 18 percent for AEZ 16. (see Figure 8.6 for spatial distribution of AEZ 16; Color legend in Figure 8.18).

Figure 8.14. (above) Inland valley bottoms delineated using the espoused methodology for Landsat path:197 and row 52. There was 8.57 percent (269,000 ha.) valley bottoms (as shown in this plate) in total area of 3.14 Mha. The valley bottom area as a percent of total area was: 7.7 for AEZ 2 and 9.1 percent for AEZ 1. (see Figure 8.6 for spatial distribution of AEZ 2 and AEZ 7 within this image; Color legend in Figure 8.18).

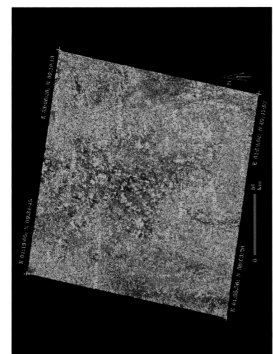

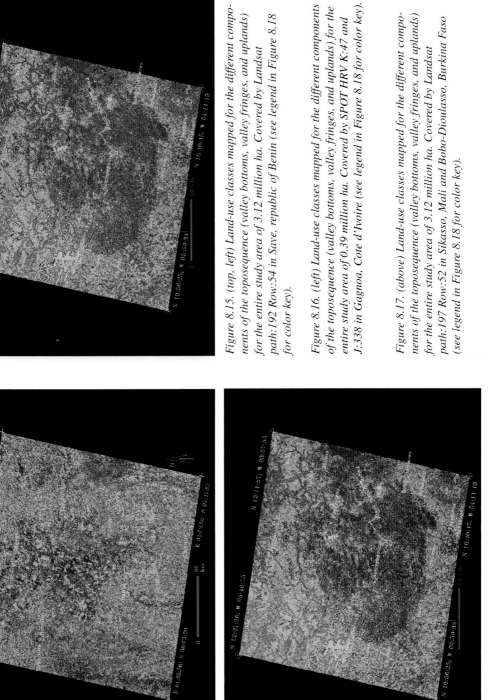

Figure 8.15. (top, left) Land-use classes mapped for the different components of the toposequence (valley bottoms, valley fringes, and uplands) for the entire study area of 3.12 million ha. Covered by Landsat path:192 Row:54 in Save, republic of Benin (see legend in Figure 8.18 for color key).

Figure 8.16. (left) Land-use classes mapped for the different components of the toposequence (valley bottoms, valley fringes, and uplands) for the entire study area of 0.39 million ha. Covered by SPOT HRV K:47 and J:338 in Gagnoa, Cote d'Ivoire (see legend in Figure 8.18 for color key).

Figure 8.17. (above) Land-use classes mapped for the different components of the toposequence (valley bottoms, valley fringes, and uplands) for the entire study area of 3.12 million ha. Covered by Landsat path:197 Row:52 in Sikasso, Mali and Bobo-Dioulasso, Burkina Faso (see legend in Figure 8.18 for color key).

Figure 8.20. Location of potential benchmark research sites for technology development research activities in the study area covered by Landsat TM path: 197 and Row:52 (Sikasso, Mali; Bobo-Dioulasso, Burkina Faso). These locations were selected by integrating remote sensing, GIS, ground-truth, and expert knowledge data in a GIS model.

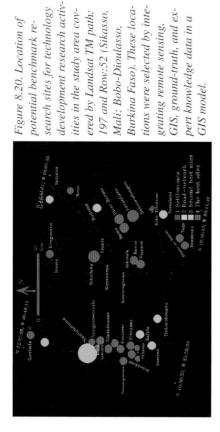

Figure 8.18. Legend of color keys for Figures 8.12–8.17.

1. Significant farms, uplands
2. Scattered farms, uplands
3. Savanna vegetation, uplands
4. Wetland/Marshland, uplands
5. Dense forests, uplands
6. V. dense forests, uplands
7. Significant farms, fringes
8. Scattered farms, fringes
9. Insignificant farms, fringes
10. Significant farms, bottoms
11. Scattered farms, bottoms
12. Insignificant farms, bottoms
13. Water
14. Built-up areas/settlements
15. Roads
16. Barren land or desert

Figure 8.21. Study of biomass levels in the valley bottoms in summer offer a powerful means of assessing potential of valleys for dry season cultivation and in general assessment of moisture levels. The normalized difference vegetation indices (NDVI) has been widely accepted as powerful means of biomass assessment. The above NDVI based biomass levels are for Nyankadougou, South-West of Bobo-Dioulasso, Burkina Faso.

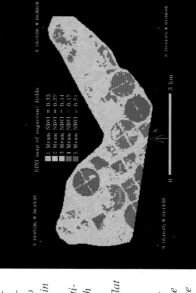

Figure 8.19. Rice cultivation in inland valleys bottoms surrounding Sikasso Mali (AEZ 2). The areas in magenta have either insignificant or no rice cultivation, but have very high potential for exploitation for rice cultivation. The flat and large valley bottoms, which are significantly flooded during rainy season and fertile soils, make these valleys ideal for rice cultivation. Special varieties that can withstand flooding are required.

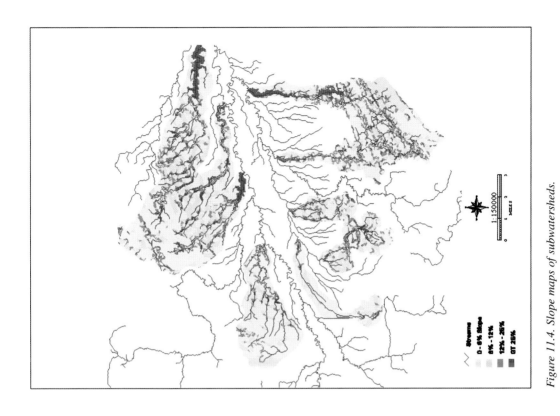

Figure 11.4. Slope maps of subwatersheds.

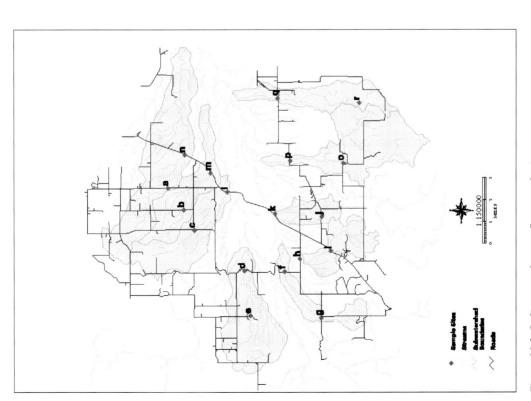

Figure 11.3. Sediment and streamflow sample site map.

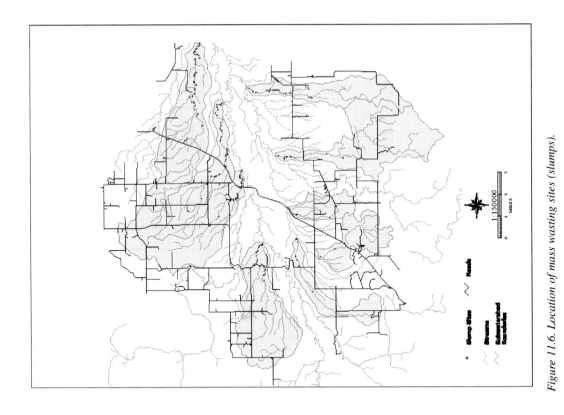

Figure 11.6. Location of mass wasting sites (slumps).

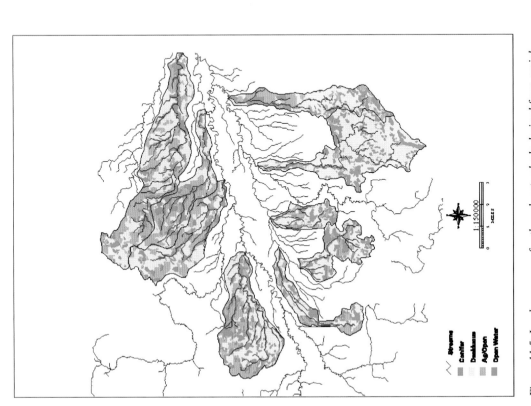

Figure 11.5. Land cover map for the subwatersheds derived from aerial photography.

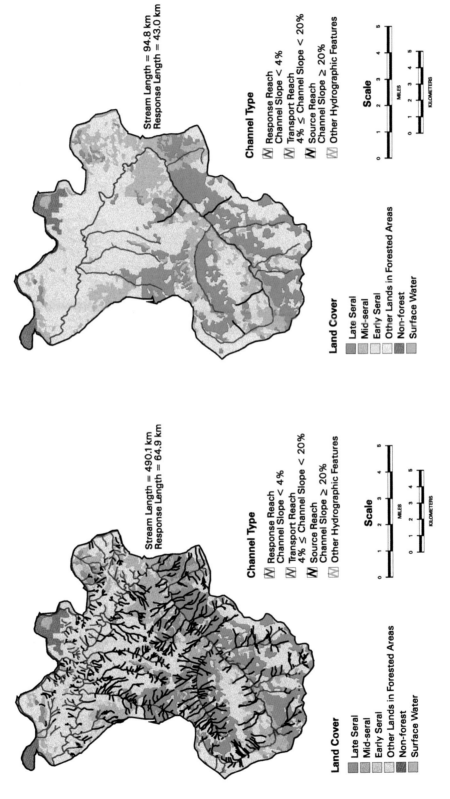

Finney Creek WAU 1:24,000-scale Hydrography

Stream Length = 490.1 km
Response Length = 64.9 km

Finney Creek WAU 1:100,000-scale Hydrography

Stream Length = 94.8 km
Response Length = 43.0 km

Channel Type
- Response Reach
  Channel Slope < 4%
- Transport Reach
  4% ≤ Channel Slope < 20%
- Source Reach
  Channel Slope ≥ 20%
- Other Hydrographic Features

Land Cover
- Late Seral
- Mid-seral
- Early Seral
- Other Lands in Forested Areas
- Non-forest
- Surface Water

Scale

MILES
0   1   2   3   4   5

KILOMETERS
0   1   2   3   4   5

Figure 14.1 (a,b). Comparison of 1:24,000 (a) versus 1:100,000 (b) scale hydrography data. Note the greater stream density and 50% increase in response reach length associated with the 1:24,000 scale data.

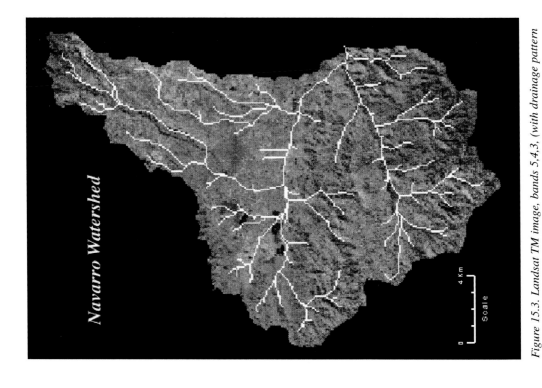

Figure 15.3. Landsat TM image, bands 5,4,3, (with drainage pattern overlay).

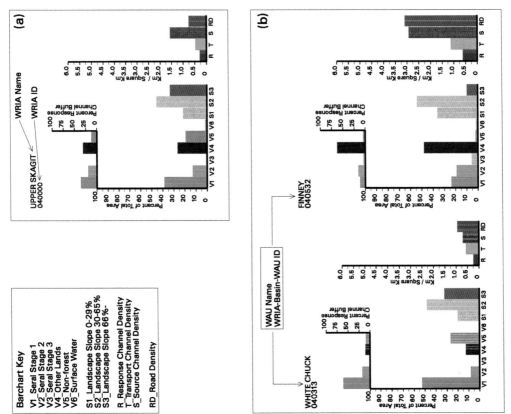

Figure 14.3 (a,b). Example WRIA bar chart (a) data presentation and WAU bar charts (b) standard outputs produced for all of western Washington State.

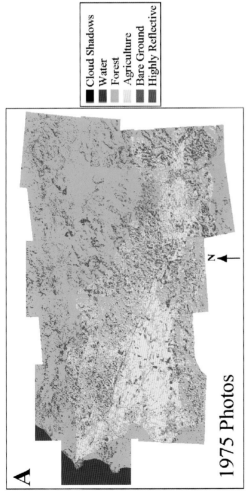

A

Cloud Shadows
Water
Forest
Agriculture
Bare Ground
Highly Reflective

N

1975 Photos

Figure 16.3A. Classified land-cover image of western Puerto Rico derived from 1975 black and white aerial photographs (Miller and Cruise, 1995).

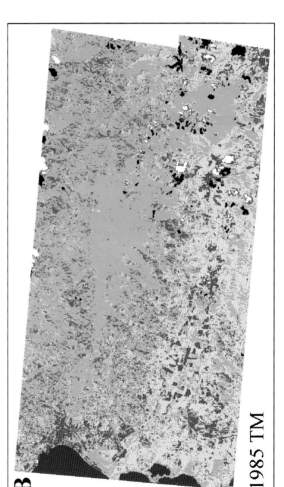

B

1985 TM

Figure 16.3B. Classified land-cover image of western Puerto Rico derived from 1985 Landsat Thematic Mapper image (Miller and Cruise, 1995).

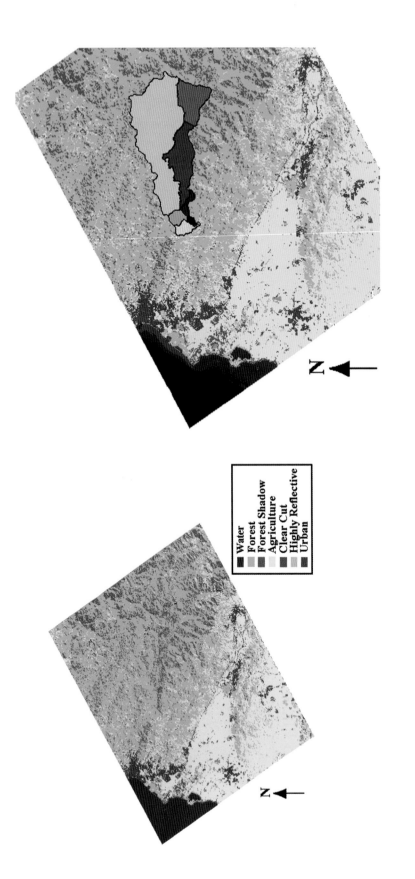

Figure 16.5. Computational zones of Rio Rosario subbasin overlain on classified 1990 CAMS image (Cruise and Miller, 1993).

Figure 16.4. Classified land-cover image of western Puerto Rico derived from 1990 CAMS data (Cruise and Miller, 1993).

Water
Forest
Forest Shadow
Agriculture
Clear Cut
Highly Reflective
Urban

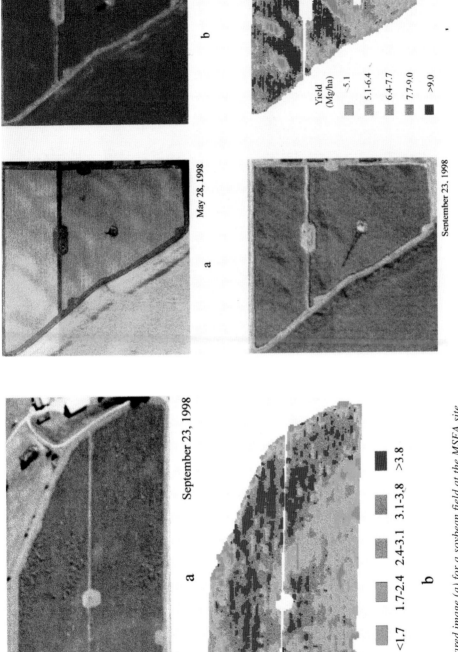

Figure 19.5. Digital infrared images and a yield map for a cornfield at the MSEA site. Note that green vegetation is red or magenta in color on the original and bare soils or areas with poor canopy development are light-toned and blue-green in color.

Figure 19.4. Digital infrared image (a) for a soybean field at the MSEA site (1998). Green areas in the original infrared image are senescent soybeans and red-brown areas are Canada Thistle, Giant Ragweed, and Johnson Grass weeds. Image (b) is an on-the-go yield monitor results from the same field.

SEP 15 Image
5 Bands – DEM
Classification

a

b

Figure 19.7a & b. Figure 19.7(a). Computer categorized images of crop, soils, and hydrological conditions. These images have more than 30 classes of types based on the light reflectance in the visible, near infrared and middle infrared and use of the Digital Elevation Model (DEM). Figure 19.7(b). False color composite of the MSEA site during senescence. Note dark spots in the top left field that are perennial weeds that stand out against the light-toned senescent soybeans.

7.7% to 18% of geographic area have great economic significance with rich potential for agricultural and ecological exploitation [see an example in Figure 8.19 (see color section) for rice cultivation] and preservation and hence deserve to be mapped as separate land units. The land-use classification resulted in estimates of areas covered by savannas, forests, farmlands, wetlands, and other land-uses (water, settlements, road networks, and barren areas) apart from estimating percentage areas cultivated and the the percentage areas still available for exploitation.

The area of inland valleys was significantly higher in wetter agroecological zones compared to drier zones. For example, valley bottoms in humid forest (AEZ 16) were 18% (Figure 8.13, color section; Table 8.7) and in derived savanna (AEZ 7) were 10.2% (Figure 8.12, color section; Table 8.5) compared to much lower areas in the northern Guinea savanna (AEZ 1) with 9.1% (Figure 8.14, color section; Table 8.9) and southern Guinea savanna (AEZ 2) with 7.7% (Figure 8.14, color section; Table 8.9) and 7.9% (Figure 8.12, color section and Table 8.5). In each study area there was a strong relationship between inland valley and upland cultivation (Tables 5, 7, and 9).

There was no clear relationship between the degree of cultivation and the agroecological zones. One of the hypotheses (Izac et al., 1991) was that drier zones will have a greater degree of valley bottom utilization than the wetter zones as moisture availability in the valley bottoms is more crucial in drier zones than wetter zones. This hypothesis is not a practical reality as other factors play an important role in inland valley utilization. Availability of reliable rainwater supply during wet season, relatively better access to technologies, higher population densities, and higher densities of road networks and socioeconomic factors facilitated utilization of 20.6% (Table 8.7) of valley bottoms, mostly for rice cultivation, in Gagnoa study area. In contrast, the percentage utilization of valley bottoms were, generally, much lesser in drier agroecological zones of Save (Table 8.5) and Sikasso (Table 8.9) due to relative absence of above-mentioned factors.

Digital georeferenced databases generated from the study consisted of the spatial distribution of inland valleys, land-use characteristics of inland valleys and their uplands, and major settlements and road-networks. Other spatial data layers like GPS data, ground-truth data, and expert knowledge databases were integrated with remote sensing derived digital databases into a Geographic Information Systems (GIS) framework. Spatial modeling of these data layers resulted in deriving potential benchmark research areas for technology development research activities (Figure 8.20, color section).

REFERENCES

Acres, B.D., R.B. Blair, R.M. King, Lawton, A.J.B. Mitchell, and L.J. Rackham, 1985. African dambos: Their distribution, characteristics and use. *Zeitschrift für Geomorphologie*, Supplementband 52, Berlin-Stuttgart, pp. 63–86.

Albergel, J., J.M. Lamachère, B. Lidon, A.I. Mokadem, and W. van Driel, (Eds.), 1993. Mise en valeur des bas-fonds au Sahel. Typologie, Fonctionnement hydrologique, Potentialités agricoles. Rapport final d'un projet CORAF-R3S, Ouagadougou, Burkina Faso, CIEH.

Anderson, J.R., E.E. Hardy, J.T. Roach, and R.E. Witmer, 1976. A Land-Use and Land-Cover Classification System for Use with Remote Sensor Data. Geological Survey Professional Paper 964, U.S. Government Printing Office, Washington, D.C.

Andriesse, W. 1986. Area and distribution. In A.S.R. Juo and J.A. Lowe. *Proceedings of an International Conference on Wetlands Utilization for Rice Production in Sub-Saharan Africa*, 4–8 November 1985, IITA, Ibadan, Nigeria, pp. 15–30.

Andriesse, W., and L. Fresco, 1991. A Characterization of rice-growing environments in West Africa. *Agriculture, Ecosystems, and Environment*, 33:377–395.

Becker, L., and R. Diallo, 1992. Characterization and Classification of Rice Agroecosystems in Côte d'Ivoire. WARDA/ADRAO, Bouaké, Côte d'Ivoire.

Bertrand, R., 1973. Contribution à l'étude hydrologique, pédologique et agronomique des sols gris sableux hydromorphes de Casamance (Sénégal). *L'Agronomie Tropical*, 28(12):1145–1192.

ERDAS, 1992. Field guide. ERDAS Inc., 2801 Buford Highway, Suite 300, Atlanta, GA.

FAO/UNESCO, 1974. Soil Map of the World, 1:5,000,000 Vol. VI. Africa. UNESCO, Paris.

FAO/UNESCO, 1977. Soil Map of the World. Revised Legend. FAO World Soil Resources Rep. 60, Rome.

Hekstra, P., W. Andriesse, G. Bus, and C.A. de Vries, 1983. Wetland Utilization Research Project, West Africa. Phase I. The Inventory: Vol. I, Main Report; Vol. II, The Physical Aspects; Vol. III, The Agronomic, Economic and Sociological Aspects; Vol. IV, The Maps. WURP-Report, ILRI/STIBOKA, Wageningen, the Netherlands.

Ingram, J., 1991. Wetlands in Drylands: The Agroecology of Savanna Systems in Africa. Part 2: Soil and Water Processes. International Institute for Environment and Development (IIED), Drylands Programme, London, UK.

Izac, A. -M. N., M.J. Swift, and W. Andriesse, 1991. A Strategy for Inland Valley Agroecosystem Research in West and Central Africa. RCMP Research Monograph No.5, Resource and Crop Management Program, IITA, Ibadan, Nigeria.

Izac, A.-M., S.S. Jagtap, A.I. Mokadem, and P.S. Thenkabail (in preparation). Agroecosystems characterization in International Agricultural Research: The case of inland valley systems in West and Central Africa. *Agriculture, Ecosystems, and Environment.*

Juo, A.S.R., and Lowe, J.A., 1986, The Wetlands and Rice in Sub-Saharan African. The International Institute of Tropical Agriculture, Ibadan, Nigeria.

Kilian, J. 1972. Contribution à l'Etude des Aptitudes des Sols à la Riziculture des Bas-Fonds sans Aménagements dans le Nord Dahomey (Bas-Fonds de Sirarou et de Bagou*). L'Agronomie Tropical* 27 (3): 321–357.

Kolawole, A., 1991. Wetlands in Drylands: The Agroecology of Savanna Systems in Africa. Part 3a: Economics and Management of Fadama in Northern Nigeria. International Institute for Environment and Development (IIED), Drylands Programme, London, UK.

Mäckel, R., 1985. Dambos and related landforms in Africa: An example for the ecological approach to tropical geomorphology. *Zeitschrift für Geomorphologie*, Supplementband 52, Berlin-Stuttgart, 1–24.

Mokadem, A.I., 1992. Apports de la Télédétection à l'étude des bas-fonds. Cas d'étude au Sierra Leone. Faculté des Sciences Agronomiques, Dép. du Génie Rural, U.E.R. Hydraulique Agricole, Laboratoire de Télédétection et d'Agrohydrologie, Gembloux, Belgique.

Oosterbaan, R.J., H.A. Gunnewig, and A. Huizing, 1987. In Water Control for Rice Cultivation in Small Valleys of West Africa, pp. 30–49. Annual Report 1986, ILRI, Wageningen, the Netherlands.

Rattray, J.M., R.M.M. Cormack, and R.R. Staples, 1953. The Vlei areas of Southern Rhodesia and their uses. *Rhodesian Agricultural Journal*, 50: 465–483.

Raunet, M., 1982. Les Bas-Fonds en Afrique et à Madagascar. Formation, Caractères Morphopédologiques, Hydrologie, Aptitudes Agricoles. IRAT, Service de pédologie, Montpellier, France.

Savvides, L., 1981. Guidelines to Improve Valley Swamps (Fadamas). Bida Agric. Dev. Project, Nigeria.

Thenkabail, P.S., and C. Nolte, 1996. Capabilities of Landsat–5 Thematic Mapper ™ data in regional mapping and characterization of inland valley agroecosystems in West Africa. *International Journal of Remote Sensing*, 17(8): 1505–1538.

Thenkabail, P.S. and C. Nolte, 1995a. Mapping and Characterizing Inland Valley Agroecosystems of West and Central Africa: A Methodology Integrating Remote Sensing, Global Positioning System, and Ground-Truth Data in a Geographic Information Systems Framework. RCMD Monograph No.16, IITA, Ibadan, Nigeria.

Thenkabail, P.S. and C. Nolte, 1995b. Level II Characterization of Inland Valley Agroecosystems in Save, Parakou, Bassila, Bante, and Tchaourou Regions of the Republic of Benin through Integration of Remote Sensing, Geographic Information Systems and Global Positioning Systems. Inland Valley Characterization Report No.1. Resource and Crop Management Division, IITA, Ibadan, Nigeria.

Thenkabail, P.S. and C. Nolte, 1995c. Regional Characterization of Inland Valley Agroecosystems in Gagnoa (Côte 'Ivoire) through Integration of Remote Sensing, Geographic Information Systems and Global Positioning Systems. Inland Valley Characterization Report No.2. Resource and Crop Management Division, IITA, Ibadan, Nigeria.

Thenkabail, P.S. and C. Nolte, 1995d. Regional Characterization of Inland Valley Agroecosystems in Sikasso (Mali) and Bobo-Dioulasso (Burkina Faso) through Integration of Remote Sensing, Geographic Information Systems and global positioning systems. Inland Valley Characterization Report No.3. Resource and Crop Management Division, IITA, Ibadan, Nigeria.

Turner, B., 1977. The Fadama Lands of Central Northern Nigeria: Their Classification, Spatial Variation, Present and Potential Use. Ph.D. Dissertation, University of London, UK.

Turner, B., 1985. The classification and distribution of fadamas in central Northern Nigeria. Zeitschrift für Geomorphologie, Supplementband 52, Berlin-Stuttgart, 87–114.

Windmeijer, P.N., and W. Andriesse, Eds. 1993. Inland Valleys in West Africa. An Agro-Ecological Characterization of Rice-Growing Environments. International Institute for Land Reclamation and Improvement (ILRI) Publ. 52, ILRI, Wageningen, the Netherlands.

CHAPTER 9

Watershed Characterization by GIS for Low Flow Prediction

Glenn S. Warner, Andrés R. García-Martinó, Frederick N. Scatena, and Daniel L. Civco

INTRODUCTION

A GIS was used to determine watershed characteristics for development of multiple regression equations for prediction of low flow in streams in the eastern part of Puerto Rico. A preliminary selection of 45 gauged watersheds was based on a general Life Zone Map, and vector coverage of precipitation and temperature regions typical of the Caribbean National Forest. The number of watersheds was reduced to 19 by overlaying land use coverage in a GIS and selecting those watersheds having permanent forest cover and available data including digital elevation models (DEMs). A total of 53 initial watershed parameters (grouped into geology, soils, geomorphology, stream network, relief, and climate) were determined by GIS for each of the 19 watersheds. The number of parameters was reduced to 13 through a two-step process using criteria that included correlation among the parameters, correlation with low flows, colinearity, significance of coefficients, and logic. Multiple regression analyses were performed to predict the 7 day–10 yr, 30 day–10 yr, and 7 day–2 yr low flows using combinations of the 13 parameters. Selection of the final regression model for each low flow was based on five statistical tests. Four parameters (drainage density, weighted mean slope, an aspect parameter, and ratio of tributaries to main channel length) were found to be the best predictors, resulting in R^2 from 0.96 to 0.97 for the three low flow variables. A non-GIS model was also developed with easier to measure parameters, but resulted in a higher SE than the GIS-based models.

BACKGROUND

Due to the complexity involved in the generation of low flow in streams, few if any physically based models have been developed for prediction of low flows. Instead, measured low flows are usually regressed against physical parameters that characterize the watersheds involved. Parameters used in low flow studies have varied considerably from one region or study to another, but typically include parameters that reflect basin, climatic, and cultural characteristics (Singh and Stall, 1974) that may play a role in the generation of low flow. Examples of parameters include watershed area, surficial geologic materials, watershed shape, and stream length.

A number of researchers have used GIS for watershed characterization for purposes such as modeling of water quality (Cahill et al., 1993), runoff (Luker et al., 1993), and nonpoint pollution (Robinson and Ragan, 1993). However, GIS application in low flow research is very limited. Low flow studies typically include a large number of watersheds with great differences in drainage areas, topography, geology, and other watershed characteristics. GIS provides not only a format for more efficient determination of these parameters, but also opportunities for investigation of parameters that are not easily measured by manual means and that therefore have not been used in most low flow studies. The development of automated computer programs for watershed characterization (Martz and Garbrecht, 1993; Eash, 1994) could lead to routine use of GIS in determination of parameter values for multiple regression equations.

The need for low flow prediction and the availability of stream gauge data and GIS coverages for eastern Puerto Rico provides an opportunity to study low flow in streams draining the Caribbean National Forest, also known as the Luquillo Experimental Forest (LEF). The LEF is located in the eastern end of Puerto Rico and covers an area of 11,300 ha. Elevations range from 73 m to 1,075 m above sea level. The LEF is almost completely forested and receives an average of 3864 mm of rainfall annually (García-Martinó, 1996). Water demand by municipalities in the area is putting increased pressure for water withdrawal from streams draining the mountains. The lack of information on low flow for ungauged streams provides a need for better methods of predicting low flows in these streams.

DATA AND METHODS

The GIS database included land use, soils, geology, hydrography (all in vector format), and Digital Elevation Models (DEM). All GIS coverages were obtained as $7.5' \times 7.5'$ quadrangles and georeferenced to the Universal Transverse Mercator (UTM) coordinate system. Soil and geology coverages were digitized from 1:20,000 scale maps from the USDA Soil Surveys (published in the 1970s) and the U.S. Geological Survey (USGS), respectively. Unconsolidated material formations as small as 7,000 m^2 are depicted in the geology coverages. The land use data (produced by the Puerto Rico Dept. of Transportation for 1977) were divided using first degree units based on Anderson et al. (1976) and highly subdivided to specific categories. The stream network was obtained from the USGS as Digital Line Graphs (DLG), showing rivers and major tributaries. The stream network shown in the DLGs was virtually the same as that shown on USGS topographic maps.

We obtained DEMs from the USGS in two different formats: 30 m by 30 m DEMs (USGS, 1987) for the eastern region of the island, specifically for El Yunque, Fajardo, Humacao, Naguabo, Juncos, and Fajardo quadrangles; and a DEM produced by the Defense Mapping Agency (DMA) for the rest of the island. The DEM from the DMA has a resolution of 3×3 arc seconds, which in Puerto Rico is equivalent to an approximate 90 m x 90 m resolution. All DEMs are in UTM coordinates with elevations in meters.

A typical low flow study involves watershed selection, frequency analysis, watersheds characterization, and multiple regression analysis. Since we were interested in the humid montane regions of Puerto Rico, we limited our investigations to watersheds either within the LEF or with rainfall and land use characteristics similar to those of the LEF. The frequency analysis for low flow of the gauged streams and the regression analysis followed standard procedures. The watershed characterization, however, was not limited to commonly used parameters that could be measured by hand, but initially included a large number of potentially important factors that either alone or in combination with other factors might be significant hydrologically. The number and complexity of the parameters considered were only feasible when using GIS, since the effort to evaluate each parameter for each watershed by hand would have been overwhelming.

To select as many representative watersheds as possible, we followed a scheme that first digitized the Life Zones Map for Puerto Rico produced by Ewel and Whitmore (1973) using the Holdridge classification system. We next performed a broad, general selection of USGS stream gauges by a visual inspection of the Life Zones Map and major drainage networks of the island.

We looked first for stations draining the LEF, and second, for stations with temperature and precipitation characteristics similar to those in the LEF. The first GIS analysis consisted of overlaying the watersheds of 45 stations (continuous and partial) with Life Zone and land use coverages. The watershed boundary for each of the 45 stations was drawn on USGS topographic maps with a 1:20,000 scale, and digitized.

We reduced the number of stations to 19 (in three different regions of the island) based on land use and life zone classification, followed by GIS application of the geology, soils, stream network and DEM coverages. We initially chose a large number of parameters based on their potential to influence storm runoff and groundwater dynamics. Many parameters such as soil water-holding capacity, liquid limit, texture, and depth to bedrock did not vary significantly across watersheds and were dropped. A total of 53 parameters (Table 9.1) covering six groups were determined by GIS to vary across the 19 watersheds and were chosen for further analysis. We found that GIS has several advantages over traditional methods: (1) a large number of parameters can be evaluated for numerous watersheds, (2) parameters not easily determined by hand such as percent of area having slopes or aspects less than various thresholds can be calculated, and (3) area-weighted parameters such as the weighted mean elevation within the watershed can be determined easily.

The data for most of the partial stations consist of six or seven low flow measurements taken over a four year period. Some low flow values were obtained from Santiago (1992). Measured low flows for other partial and three continuous stations with short records were correlated with concurrent discharges from nearby watersheds with similar hydrologic characteristics as recommended by Riggs (1972). Daily low flow measurements were randomly selected for each of these stations and correlated with concurrent low flows from nearby continuous stations. All correlations were significant ($\alpha = 0.05$).

We developed two types of regression models: GIS-based models (for 7d–10yr, 30d–10yr, and 7d–2yr low flows) that considered all 13 parameters and a non-GIS model (7d–10yr only) that only considered parameters that could be easily measured without a GIS. A model can be statistically strong, but if the parameters are difficult to measure, then the model may be useless. Although GIS technology is dramatically closing the gap between easy and hard to measure parameters, GIS is not always available. We selected the best models of each type using a number of statistics given by Helsel and Hirsch (1992). Parameters considered in each model and details of the statistical tests are given in García-Martinó (1996).

RESULTS AND DISCUSSION

Four parameters consistently produced the best overall results for the GIS-based models: DD is total length of perennial channels per unit area (m/m^2), SLP is weighted mean slope in the watershed ($°$), PA90 is percent of total area of watershed with aspects between 0–90°, and CHATRI is ratio of total tributary length to length of main channel. Although watershed area does not appear directly in the equations, it is highly correlated with other parameters. The GIS models (one for each low flow) are:

$$\text{Log7d–10yr} = 1.78 - 1331.61(\text{DD}) - 0.0316(\text{SLP}) + 0.0146(\text{PA90}) + 0.3126(\text{CHATRI})$$
$$R_a^2 = 0.97 \tag{1}$$

$$\text{Log30d–10yr} = 1.60 - 1098.26(\text{DD}) - 0.0301(\text{SLP}) + 0.0160(\text{PA90}) + 0.2957(\text{CHATRI})$$
$$R_a^2 = 0.96 \tag{2}$$

Table 9.1. Definitions of Watershed Parameters Used in Correlation and Regression Analyses

CLIMATE BASED PARAMETERS:
  MMYR = mean annual rainfall (mm/yr)
  10YRCDNR = number of consecutive days with no rain with recurrence interval of 10 years
  M#DNR = average number of days per year with no rain
  MMCDNR = maximum consecutive days with no rain

GEOLOGY BASED PARAMETERS:
  UNCON and QA = area and % of area of watershed classified as unconsolidated material, respectively
  LSEU = length of perennial streams in direct contact with unconsolidated material
  INTRU = % of area classified as intrusive igneous rock with low water bearing capacities
  STRAT = % of area classified as alluvial deposits and sedimentary formations (sandstone)

SOIL BASED PARAMETERS:
  PLASTI = area-weighted mean plasticity index for watershed
  PERM = area-weighted mean permeability

STREAM NETWORK BASED PARAMETERS:
  PERC = total length of perennial channels (m)
  CHAN = total length of main channel (m)
  TRIB = total length of tributaries to main channel (m)
  DD = total length of perennial channels per unit area (m/m^2)
  SD = number of perennial channels per unit area ($\#/m^2$)
  TRIPER = ratio of total length of tributaries to total length of perennial channels
  CHATRI = ratio of total length of tributaries to total length of main channel
  CHANSLP = slope of main channel between 15% and 85% points of main channel (°)
  CHSLP = (CHANSLP/CHAN)/100
  SINUO = total length of main channel divided by length of watershed

WATERSHED MORPHOLOGY BASED PARAMETERS:
  AREA = watershed area (ha)
  SLP = area-weighted slope of watershed (°)
  PERIM = length of perimeter of watershed boundary (m)
  COMP = ratio of PERIM to circumference of a circle with area = AREA
  FORM = AREA/(watershed length^2)
  ELONG = diameter of circle with area = AREA divided by watershed length
  AREA4 and A4, AREA7 and A7, AREA10 and 10 = areas and % of AREA with slopes < 4, 7 and 10°
  MSLP100, ASLP100, MSLP200, ASLP200 = maximum and mean slope within 100 or 200 m buffers of perennial streams (°)
  ASPECT = area-weighted aspect (°)
  AREA90 = total area of watershed with aspect between 0 and 90(°)
  PA90 = % of AREA with aspects between 0 and 90°
  ASP100 and ASP200 = mean area-weighted aspect inside 100 and 200 m buffers along perennial streams

WATERSHED RELIEF BASED PARAMETERS:
  EL = area-weighted mean elevation
  ELRANGE = maximum minus minimum elevation in watershed (m)
  RELIEF = ELRANGE divided by length of watershed
  RERA = RELIEF*ELRANGE
  SLORAT = CHANSLP/SLP
  HYP1 = % change between 0.25 and 0.75 points of hypsometric curve of watershed
  HYP2 = HYP1/AREA
  HYP3 = HYP1/(% change between 0.50 and 0.75 points of hypsometric curve)

$$\text{Log7d-2yr} = 1.70-1157.49(DD)-0.0248(SLP)+0.0121(PA90)+0.3017(CHATRI)$$
$$R_a^2 = 0.97 \tag{3}$$

where R_a^2 is the adjusted R^2 (Velleman, 1992).The best non-GIS model consisted of four parameters that could be relatively easily measured by manual methods for prediction of the 7d–10yr flows:

$$\text{Log7d-10yr} = 0.1827-0.0824(CHANSLP)+2.24(FORM)+0.0012(ELRANGE)-882.51(DD)$$
$$R_a^2 = 0.94 \tag{4}$$

where: CHANSLP is slope between 15% and 85% points of main channel, FORM is area divided by square of watershed length, ELRANGE is maximum minus minimum elevation in the watershed, and DD is same as for GIS-based models above.

The error in a model is the product of chance variation and variation from unaccounted differences in watershed characteristics. The objective is to decrease the variation due to the latter. The variation in Y due to each variable (R^2) is very similar in each GIS model (Table 9.2). The strongest variable is DD, contributing over 53% of the variation in Y in each model. The strongest parameter in the non-GIS model was also DD (Table 9.2), which evidently explains a high degree of the subsurface and groundwater dynamics.

The best method of determining the accuracy of the non-GIS model compared to the GIS model is the standard error of the estimate (SE). The SEs varied from 11.5% to 13.8% for the GIS models compared to 19.5% for the non-GIS model (Table 9.2). Values for SE over 100% are common (Riggs, 1973). One of the most accurate low flow models was developed for West Virginia by Chang and Boyer (1977) and had an SE of 31%. Since SE tends to be underestimated for small samples, an adjusted SE (SE_a) as suggested by Ezekiel and Fox (1959) was calculated by the following equation:

$$SE_a = [N/(N-M) * SS_e]^{0.5} \tag{5}$$

where N is the sample size, M is the number of coefficients in the model including the intercept, and SS_e is the sum of squares of the error. The SE_a's for the GIS models shown in Table 9.2 are good, even when compared with the nonadjusted SE more commonly found in low flow research. Although the variation in Y covered by the non-GIS model is relatively high ($R^2 = 0.94$), the SE_a of 82.8% is well above the SE_a of the GIS models. It is still a good model when compared to most low flow research in the U.S. A direct comparison of the GIS model vs. non-GIS model can be made for the 7d-10yr which shows an increase in SE_a from 58.7% (GIS) to 82.8% (non-GIS) that can be attributed to the ability to include additional and more complex parameters by GIS.

The future availability of greater resolution DEMs for areas outside the LEF will significantly improve the analysis. Greater resolutions will permit more detailed and precise watershed characterization, especially when dealing with slopes and aspect. Also, it will be possible to extend the stream network to streams not included in a USGS topographic map. GIS coverages are available for the whole island. These or similar models can be developed and/or adapted to the unique conditions for each region of the island in order to characterize the watersheds and develop better water management plans. On the other hand, the accuracy of the selected models is relatively high, and its application to humid montane regions in Puerto Rico and adjacent areas should provide opportunities for more intensive analysis and greater confidence for low flow predictions.

Table 9.2. Parameters and Comparative Statistics for Selected GIS and Non-GIS Models

GIS Models Parameters	Model C-3 (7 day–10 yr)			Model D-4 (30 day–10 yr)			Model E-3 (7 day–2 yr)		
	R_a^2	SE[1]	SE_a[2]	R_a^2	SE	SE_a	R_a^2	SE	SE_a
DD	0.59	45.68	193.81	0.53	44.67	189.51	0.54	43.19	183.25
DD + CHATRI	0.91	21.84	92.68	0.88	23.41	99.32	0.92	18.11	76.85
DD + CHATRI + PA90	0.96	15.71	66.65	0.95	15.59	66.15	0.96	12.83	54.43
DD + CHATRI + PA90 + SLP	0.97	13.84	58.72	0.96	13.94	59.15	0.97	11.47	48.64

Non-GIS Models Parameters	Model P-1 (7 day-10 yr)		
	R_a^2	SE[1]	SE_a
DD	0.59	45.68	193.81
DD + ELRANGE	0.74	37.55	159.33
DD + ELRANGE + FORM	0.88	26.84	113.88
DD + ELRANGE + FORM + CHANSLP	0.94	19.52	82.81

[1]Standard error.
[2]Adjusted standard error.

SUMMARY AND CONCLUSIONS

The application of GIS for watershed characterization for low flow prediction provides the opportunity to evaluate large numbers of parameters and to evaluate parameters which are difficult to determine by hand methods. Unique and highly qualitative parameters could be readily measured using GIS. The GIS-based models produced better estimates of low flows than use of a non-GIS model developed for gauged watersheds in humid montane regions of Puerto Rico. The adjusted standard error of the estimate (SE_a) of the GIS-based model for the 7d–10yr flow was 58.7% compared to 82.8% for the non-GIS model. These SE_a's are less than the error of most low flow research performed in the U.S. The most significant parameters in the GIS model for each of the three low flows were drainage density (DD), the ratio of the length of the tributaries to the length of the main channel (CHATRI), the percent facing northeast (PA90), and the mean slope in the watershed (SLP).

ACKNOWLEDGMENT

Funding for this study was provided by the USDA Forest Service through the Luquillo Experimental Forest, Rio Piedras, Puerto Rico. Published as Scientific Contribution No. 1700 of the Storrs Agricultural Experiment Station.

REFERENCES

Anderson, J.A., E.E. Hardy, J.T. Roach, and R.E. Witmer, 1976. A Land Use and Land Cover Classification System for Use with Remote Sensing Data. U.S. Geol. Survey Prof. Paper 954.

Cahill, T.H., J. McGuire, and C. Smith, 1993. Hydrologic and water quality modeling with geographic information systems. In *Geographic Information Systems and Water Resources*, J.M. Harlin and K.J. Lanfear, Ed., American Water Resources Association, Mobile, AL, pp. 313–317.

Chang, M., and D.G. Boyer, 1977. Estimates of low flows using watershed and climatic parameters. *Water Resources Research*, 13(6): 997–1001.

Eash, D.A., 1994. A geographic information system procedure to quantify drainage-basin characteristics. *Water Resources Bulletin*, 30(1):1–17.

Ewel, J.J., and J.L. Whitmore, 1973. The Ecological Life Zones of Puerto Rico and the U.S. Virgin Islands. USDA Forest Services. Resource Paper. ITF–18. Institute. of Tropical Forestry, Rio Piedras, PR.

Ezekiel, M., and K.A. Fox, 1959. *Methods of Correlation and Regression Analysis*. John Wiley and Sons, New York, NY.

García-Martinó, A.R., 1996. "Use of GIS for Low Flow Prediction in Humid Montane Regions in Eastern Puerto Rico." Unpublished M.S. Thesis. University of Connecticut, Storrs.

HEC, 1992. HEC-FFA: Flood Frequency Analysis User's Guide. U.S. Corps of Engineers, CPD–13.

Helsel D.R., and R.M. Hirsch, 1992. *Statistical Methods in Water Resources*. Elsevier Science Publishing, Co., New York, NY.

Luker, S., S.A. Samson, and W.W. Schroeder, 1993. Development of a GIS based hydrologic model for predicting direct runoff volumes. In *Geographic Information Systems and Water Resources*, J.M. Harlin and K.J. Lanfear, Eds., American Water Resources Association, Mobile, AL, pp. 303–312.

Martz, L.W., and J. Garbrecht, 1993. Automated extraction of drainage network and watershed data from digital elevation models. *Water Resources Bulletin*, 29(6):901–916.

Riggs, H.C., 1973. Regional Analyses of Streamflow Characteristics. U.S. Geological Survey Techniques of Water-Resources Investigations Bk. 4, Chap. B3.

Riggs, H.C., 1972. Low Flow Investigations. U.S. Geological Survey Techniques of Water-Resources Investigations Bk. 4, Chap. B1.

Riggs, H.C., 1967. Some Statistical Tools in Hydrology. U.S. Geological Survey Techniques of Water-Resources Investigations Bk. 4, Chap. A1.

Robinson, K.J., and R.M. Ragan, 1993. Geographic information system based nonpoint pollution modeling. In *Geographic Information Systems and Water Resources*, J.M. Harlin and K.J. Lanfear, Eds., American Water Resources Association., Mobile, AL, pp. 53–60.

Santiago-Rivera, L., 1992. Low-Flow Characteristics at Selected Sites on Streams in Eastern Puerto Rico. U.S. Geological Survey Water-Resources Investigations Report 92–4063.

Singh, K.P., and J.B. Stall, 1974. Hydrology of 7-day 10-year low flows. *Journal of the Hydraulics Division, ASCE*, 100:1753–1771.

U.S. Geological Survey, 1987. Digital Elevation Models Data Users Guide. National Mapping Program Tech. Intruc. Data Users Guide 5.

Velleman, P.F. 1992. Data Desk[4] ® Handbook and Statistics Guide. Data Description, Inc., Ithaca, NY.

Wadsworth, F.H. 1951. Forest management in the Luquillo Mountains. I. The setting. *Caribbean Forestry*, 12:93–114.

Water Resources Council, 1977. Guidelines for Determining Flood Flow Frequency. Bulletin #17A, Hydrology Committee.

CHAPTER 10

Evaluation of the Albemarle-Pamlico Estuarine Study Area Utilizing Population and Land Use Information

Robert E. Holman

BACKGROUND

The Albemarle-Pamlico Estuarine Study (APES) has been funding many information acquisition projects over the last five years in the areas of resource critical areas, water quality, fisheries, and human environment (Steel and Scully, 1991). Most of these projects have transferred their data over to the APES's Geographic Information System (GIS) which was created through a subcontract with the North Carolina Center for Geographic Information System (CGIA). GIS has the ability to bring together (enter, display, edit, and manipulate) data based information with digital mapping (locational attributes). At the time of this study, the Center (CGIA) had or was creating all of the needed databases. CGIA was able to combine the data layers in various ways to analyze the relationship among different layers in a visual as well as a statistical manner.

The study area encompasses approximately 23,250 square miles and includes all or portions of 37 counties in eastern North Carolina and 19 counties in Southeastern Virginia. There are six counties along the coastline, 9 counties along the sounds, and 41 cities/counties that lie in the upper drainage basin (Figure 10.1). This study also incorporates all or portions of 6 major river basins including the Chowan, Pasquotank, Lower Roanoke, Tar-Pamlico, Neuse, and White Oak (Figure 10.2). Each basin is divided into subbasins as follows: Chowan, 13; Pasquotank, 8; Lower Roanoke, 3; Tar-Pamlico, 8; Neuse, 14; and White Oak, 5.

METHOD

The analytical method was broken into three phases. Phase One was the creation of county land use maps from the existing Landsat classification scheme. These map products were sent to U.S. Fish and Wildlife and county officials to determine the accuracy of the defined land use classes. The land use maps were also used by the author during flights over the coastal and metropolitan areas to further clarify classification errors. Phase Two was to correct some of the errors in the existing classification. This was carried out by digitizing the corrections to the map products that were returned from Fish and Wildlife and county officials. The map information was also supplemented with other sources of information such as U.S. Fish and Wildlife Service—National Wetland Inventory, U.S. Forest Service—Forest Inventory and Analysis, U.S. Bureau of Census—Census of Agriculture, and U.S. Soil Conservation Service—National Resources Inventory and Hydric Soils in North Carolina counties. Phase Three was identifying correlation between

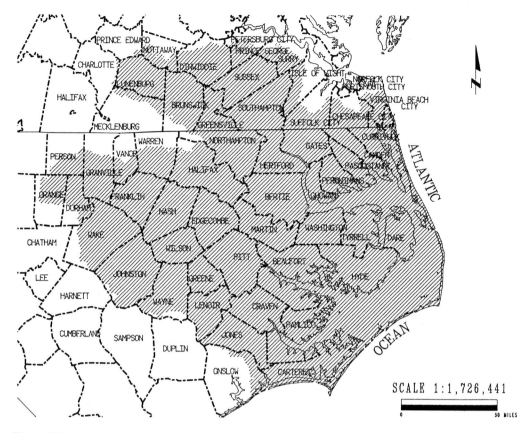

Figure 10.1. Map of study area.

different data sets such as county census population and county acreage of developed land. If a strong correlation was found, then a simple linear regression model was applied in order to predict the relationship between the two parameters. These models were used to correct some of the errors in specific land use categories.

Map Development

There were three tasks in the development of land use and population estimates for the entire APES area: (1) defining the actual drainage area; (2) having all the basin and subbasin boundaries digitized in order to determine the land use and population; and (3) correcting for errors associated with the different land uses.

First, the study area was defined as the entire drainage area of the Albemarle and Pamlico Sound system including Core and Bogue Sounds. The upper Roanoke Basin and a portion of the White Oak Basin were not included because: (1) the upper Roanoke River Basin covers approximately 8,370 square miles in Virginia/North Carolina and stretches over two-thirds the length of North Carolina, and would add one-third more area to the study area; and (2) a decision was made early in APES to have Carteret County as the furthest area south. However, due to the watershed approach in defining the study area in this project, all the subbasins in the White Oak Basin were included except the furthest one southwest that starts at Camp Lejeune. There was no compatible land use data available for this subbasin.

Figure 10.2. The six APES basins and their subbasins.

Second, all North Carolina basins and subbasins were digitized by the Research Triangle Institute and compared closely with the U.S. Geological Survey subbasins for North Carolina. Virginia subbasin information was supplied by Information Support Systems Laboratory within Virginia Polytechnic Institute and State University and was based on Soil Conservation Service (SCS) information. Due to the large number of subbasins identified by SCS in the Virginia portion of the Chowan and Pasquotank Basins, subbasins were combined to create areas of the same size range as subbasins identified in North Carolina. All subbasins were digitized from U.S. Geological Survey's 1:24,000 scale topographic maps. Specific subbasins were identified by a six number code that was broken into two-digit sets. The first two digits identified the regional basin; the second two digits identified the basin; and the third two digits identified the subbasin (Figure 10.2). Codes used in this report were the same ones adopted by the North Carolina Division of Environmental Management.

The third task was to identify the accuracy of the land use data and to develop methods to correct for the errors. Khorram and others (1992) found that with the Landsat data, the urban or built-up land use category was only 46% accurate, and the accuracy of forested wetlands was unknown. In addition, the classification of mixed pixels in the existing land use data set had to be resolved. Mixed pixels are defined as areas that could not be classified because the resolution or pixels were a mixture of many categories. The land use classification from 1987–1988 developed by Khorram will be referred to as the "Landsat" classification in this study.

Accuracy and Errors

The first task was to define which level of land use to utilize. Landsat land use classification defined 18 separate classes that can be generally broken into similar U.S. Geological Survey level I and level II groupings (Anderson and others, 1976). The land use classification was based on Landsat data which Khorram interpreted mostly as land cover (the actual extent of vegetative and other cover) and some land use (interpretation of activities taking place on the land). Interpretation of land use is much more subjective than land cover and is dependent on the knowledge of the individual interpreter.

A level II map with 18 individual classes was provided to officials of two Fish and Wildlife Refuges within the APES area for their evaluation as to land cover accuracy. The Great Dismal Swamp National Wildlife Refuge staff reviewed the Landsat land use map of the refuge. This refuge, located on the border between North Carolina and Virginia just south of Portsmouth, Virginia covers approximately 110,000 acres and is predominantly forested wetland. The staff felt there was good separation among development, agriculture, water, and forest; however, the different forest cover types had serious reliability problems. A major problem was the misclassification of wetter deciduous stands like cypress/gum and maple/gum as pine/hardwood forest. A second land cover map was sent to Mattamuskeet and Swan Quarter National Wildlife Refuges personnel for their review. These two refuges are located entirely in Hyde County, North Carolina, and Swan Quarter is adjacent to the Pamlico Sound. These refuges together cover approximately 65,700 acres and are predominantly water, wetland, and forest. The staff found quite a few areas that were referred to as mixed pixels that were actually open water and irregularly flooded brackish marshes. White Cedar stands were actually marsh impoundment areas around Lake Mattamuskeet, and pine forest was actually mixed pine/hardwood or hardwood/cypress/pine forest. In general, both refuges indicated accuracy problems with the different forest and the mixed pixel classifications.

After indicating they could not evaluate all the classifications in level II land cover maps, officials of counties in the Currituck Sound Basin south of Virginia Beach and adjacent to the Atlantic Ocean were sent, for comment, land cover maps which included the following attributes: USGS level I with 6 categories shown in color; U.S. Census TIGER files that displayed the road network, map scale of 1:100,000; and modified LUDA land use data for the urban or built-up category.

Land Use Data Analysis (LUDA) was an early GIS effort started by U.S. Geological Survey in 1975 to define the land use for the entire United States. All the photographs were manually photointerpreted. The map series consisted of 1:250,000 scale maps of North Carolina defining 37 uses based on the level II classification system. Source imagery was 1:56,000 color infrared photography and 1:80,000 black and white photography dating back to 1970. Resolution was 10 acres for the urban or built-up categories and 40 acres for the remaining classifications (Kleckner, 1981). For comparison, the Khorram classification was based on 1987–1988 Landsat satellite imagery that was semiautomatically interpreted. The county map series was at a scale of 1:100,000 with a final resolution of 1 acre.

In 1991 and 1992, the author flew along the Outer Banks and inland around the estuarine portion of the study area and over portions of Wake, Durham, and Orange Counties to verify the problems with the categories of urban or built-up, wetland, and mixed pixels. The urban or built-up class was underestimated on the 1987–1988 land use maps mainly due to forest crown cover that obscured the true land use on the ground. High spectral reflectance of bare agricultural fields was also a problem because these fields were being classified as developed areas. The problem with fields being identified as developed areas was especially evident in the Landsat scene furthest west that included the Raleigh metropolitan area. This category was a very small percentage (3.3 to

7.9%) of the overall land use of the study area but is critical in that urban or built-up land use can cause the greatest impact on natural resources of the APES area.

Problems associated with the wetland class were found to be interference from forest crown cover. Open marsh and pocosin wetlands were usually accurately defined by the 1987–1988 land use maps but closed forest canopy prevented standing water below the forest to be seen. Therefore these true wetland types were usually defined as forest.

The category of mixed pixels is a grouping the classification scheme could not identify. Flights over the coastal and metropolitan areas verified that in most cases they were a mixture of standing water and wetland vegetation. The only exception to this observation was in Pasquotank County where poorly drained agricultural land was defined as mixed pixels or wetland on the county land use map.

These land use classification problems and others were identified in a workshop the author attended to verify remotely sensed land cover data for the Coastwatch Change Analysis Program of the National Oceanic and Atmospheric Administration (Burgess and others, 1992). The problems fell into four categories: classification error, cover versus land use, categorical resolution, and change detection. Classification errors included "salt and pepper" effect of individual pixels, shadows and bare ground as urban areas, and problems with the degree of wetness during image acquisition. Cover versus land use had the inherent problem with distinguishing land uses and required ancillary data. Categorical resolution was related to the spatial resolution and improper classification. The change detection problem involved the ability to detect a change but not always the nature of the change. From the author's own observations and the results of this workshop, methods were developed to overcome some of the problems associated with remotely sensed land cover data.

Land use information used in this study was analyzed according to the Khorram and others (1992) classification system but condensed from 18 to 7 categories. Certain corrections were incorporated into some classes depending on the observed and documented error associated with each class. The LUDA data set was used to determine "developed land" because the information appeared to be closer to the actual extent and location than the original 1987–1988 Landsat data set. Corrected Landsat built-up areas on the maps returned by county officials were found to have a high degree of correlation (R^2 of 0.9) with the LUDA "developed" category. A linear regression model was used to predict built-up land from the LUDA data. U.S. Fish and Wildlife's National Wetland Inventory (NWI) data were used as a reliable source of wetland acreage for the coastal plains of North Carolina (Wilen, 1990 and Burgess and others, 1992). Wetland acres for twelve of the coastal counties was provided by Kevin Morehead of the Savannah River Ecology Laboratory. The same procedure used to correct built-up areas was used to reconcile the Landsat wetland category with the NWI acreage. A high correlation resulted with a R-squared of 0.9 and a simple linear regression model was used to predict wetland acres from the NWI values. The mixed pixel unidentified category was determined by the U.S. Fish and Wildlife personnel and two overflights of the APES area to be predominantly wetland in nature. The mixed pixel figures were incorporated into the wetland classification.

RESULTS

Land Use

The entire APES area land use classification is based on a modified U.S. Geological Survey's level I classification scheme. One fact to keep in mind is that the water class is not a true *land* use but is a very important classification. There were seven classes with the following percentages: "urban" 4.8%, "agriculture" 28.2%, "forest" 28.4%, "water" 14.6%, "wetland" 20.5%, "shrub

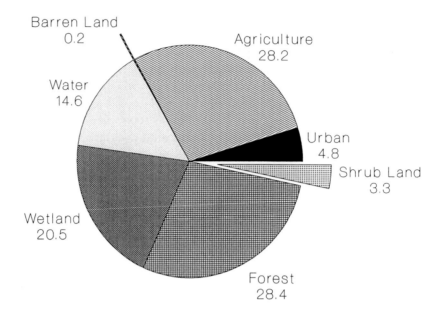

Modified 1987-88 Data

Figure 10.3. APES 1990 land use/land cover.

land" 3.3%, and "barren land" 0.2% (Figure 10.3). In general, the study area is rural in nature with less than 5% of the total area developed. More than 55% of the total APES acreage came from the categories of agriculture and forest.

Population

The population of the study area was almost 2 million people (Figure 10.4). About 51.2% reside in the Neuse Basin, which occupies only 26.8% of the land area. Population density for the basins ranged from 163.1 persons/square mile in the Neuse to 39.9 persons/square mile in the Chowan. Since the population density of all but two basins fell below the U.S average of 69 persons/square mile, most of the study area can be characterized as nonmetropolitan (U.S. Bureau of Census 1960, 1970, 1980, 1990).

Population Versus Developed Land

When the subbasins with the highest number of persons/square mile are compared to the subbasins that have the greatest amount of developed land there appears to be a great deal of agreement. If a strong correlation exists between these two parameters then a powerful planning tool can be created to predict the amount of developed land from the existing or projected population for a certain area.

The APES area has only two comprehensive land use databases and one does not correlate well with the category of urban or built-up land. Therefore, another source of long-term land use

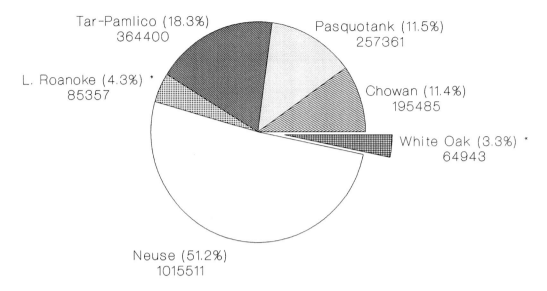

Tar-Pamlico (18.3%)
364400

Pasquotank (11.5%)
257361

L. Roanoke (4.3%) *
85357

Chowan (11.4%)
195485

White Oak (3.3%) *
64943

Neuse (51.2%)
1015511

* Only a portion of this basin

Figure 10.4. APES basin populations in 1990.

data is needed to determine if the relation between population and developed land is statistically sound. Land use data from the State of Maryland has been gathered since 1973 and has been taken as frequently as every five years during the past 15 year period (Maryland State Planning Office, 1991). The acres of total developed land were compared to the closest population census data for each county in Maryland. Three periods (1973, 1981, and 1990) had correlation with a R^2 value of greater than or equal to 0.9. A simple regression model was developed for each correlation and the results were very similar for all three periods. A population of 200,000 people was equated to between 39,000 and 43,000 acres of developed land. Since this relationship held for the Maryland data set, could the same relationship be established with the limited land use data sets in North Carolina? The earlier LUDA data set appeared to correlate well with developed land, but how could the existing Landsat accuracy for developed land be improved? Landsat land use maps of 21 counties in the APES area were sent out to county planners or other county officials for their review. Each county official was to shade in the extent of development that took place in his county during 1990 and change any land use that was not properly classified. The returned maps were digitized and new acreage for developed land was obtained for each county. Both the LUDA and the corrected Landsat land use maps were compared to the population census data in the same manner as the Maryland information. Both correlation had a R^2 value of greater than or equal to 0.8. Again, a simple linear regression model was developed for each correlation and the results were very similar for both 1970 and 1990 (Figures 10.5 and 10.6). A population of 200,000 people was equated to between 45,000 and 60,000 acres of developed land.

A statistical relationship between population census and developed land for the same time frame has been established for land area in Maryland and the APES area. The relationship is not the same for both areas and probably will vary from region to region. Based on this relationship

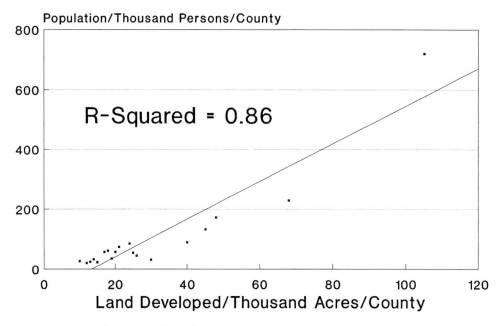

Figure 10.5. 1970 population vs 1972 land use.

the number of acres of developed land for the years 1970, 1980, and 1990 has been estimated for each of the six basins in the APES area. The resulting pattern is similar to that of population, with the Neuse Basin having the largest amount of developed land over the last 30 years. The 1990 figures show the Neuse Basin with approximately 306,000 acres of developed land and the

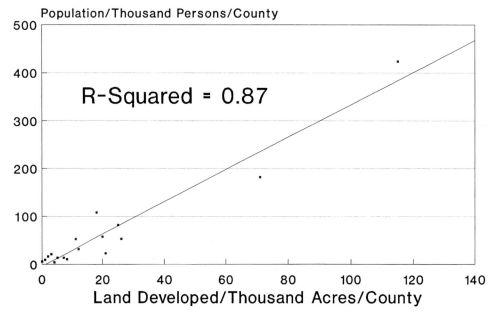

Figure 10.6. 1990 population vs 1990 land use.

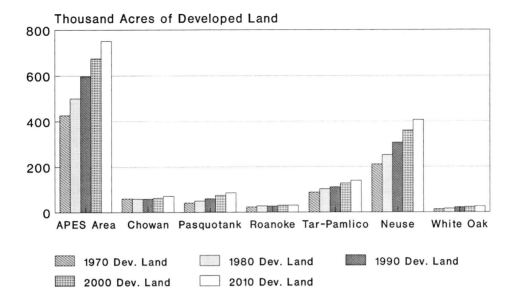

Figure 10.7. Developed land in basins.

White Oak Basin with the least at 65,000 acres of developed land. Developed land for the entire study area for 1990 was approximately 597,000 acres based on this predictive method. Projections for the year 2000 and 2010 (NCDC, 1991; VEC, 1991) have also been estimated and compared to the year 1990. Developed land for the entire study area for the year 2010 is approximately 752,000 acres based on this predictive method. The Neuse River Basin continues to have the most developed land with 407,000 acres and the White-Oak Basin has the least with 26,000 acres (Figure 10.7). The Pasquotank Basin appears to have outpaced the Chowan Basin in the amount of developed land and contains the third largest acreage behind the Neuse and Tar-Pamlico.

CONCLUSIONS

This study found that land use data from Landsat was not adequate by itself to properly identify seven land use classifications. The greatest errors appeared to be associated with the classes of built-up and wetlands. This was due to the forest crown cover obscuring the true land use of rural residential development on the ground and data sources combined with existing data sets in the form of a linear regression appeared to compensate for these two major errors.

A high correlation between population and developed land was found on a county level from Maryland, Virginia, and North Carolina data. Based on this correlation a linear regression model was developed to predict the number of acres of developed land based on the projected population for a particular county. The relation between population and developed land will not be the same for each region but this method can be a powerful tool in predicting where and how development will take place in a particular county.

REFERENCES

Anderson, J.R., E.E. Hardy, J.T. Roach, and R.E. Witmer, 1976. A Land Use and Land Cover Classification System for Use with Remote Sensor Data. U.S. Department of the Interior, U.S. Geological Survey Professional Paper 964. U.S. Geological Survey. Washington, D.C.

Burgess, W., E. Christoffers, J. Dobson, R. Ferguson, A. Frisch, P. Lade, and J. Thomas, 1992. Results of a Field Reconnaissance of Remotely Sensed Land Cover Data. Maryland Department of Natural Resources. Annapolis, MD.

Khorram, S., H. Cheshire, K. Sideralas, and Z. Nagy, 1992. Mapping and GIS Development of Land Use and Land Cover Categories for the Albemarle-Pamilco Drainage Basin. Albemarle-Pamlico Estuarine Study. Project No. 91-08. North Carolina Department of Environment, Health, and Natural Resources. Raleigh, NC.

Kleckner, R., 1981. A National Program of Land Use and Land Cover Mapping and Data Compilation. In *Planning Future Land Use*, American Society of Agronomy, Special Publication No. 42.

Maryland State Planning Office, 1991. Maryland's Land 1973–1990: A Changing Resource. Baltimore, MD.

North Carolina Data Center (NCDC), 1991. Population Projections for the Years 2000 and 2010. Raleigh, NC.

Steel, J., and M. Scully, 1991. Projects Funded by the Albemarle-Pamlico Estuarine Study. Albemarle-Pamlico Estuarine Study. Project No. 91-00. North Carolina Department of Environment, Health, and Natural Resources. Raleigh, NC.

Virginia Employment Commission (VEC), 1991. County Population Projections for the Years 2000 and 2010. Richmond, VA.

Wilen, B.O., 1990. The U.S. Fish and Wildlife Service's National Wetland Inventory. In Federal Coastal Wetland Mapping Programs: A Report by the National Ocean Pollution Policy Board's Habitat Loss and Modification Working Group. Kiraly, S.A. and F.A. Cross, Eds., U.S. Department of the Interior. Washington, D.C.

CHAPTER 11

Application of GIS and Remote Sensing for Watershed Assessment

Lloyd P. Queen, Wayne L. Wold, and Kenneth N. Brooks

INTRODUCTION

Located at the western terminus of Lake Superior, the Nemadji River Watershed covers 670 square kilometers in Minnesota (Figure 11.1). Approximately 40% of the area of the Nemadji Basin occurs in what is termed the Red Clay Area, a band of montmorillonite clays, up to 60 meters in depth. The clays, which are a result of offshore sediment deposits from glacial Lake Duluth, are highly erodible and prone to mass wasting. The high levels of sediment transported by the Nemadji River system adversely affect the designated trout streams in the basin, and deposit an estimated 525,000 metric tons of sediment annually into the Duluth/Superior harbor.

The erosion and sedimentation problems associated with the Nemadji Watershed led to the creation of the multiagency research and demonstration "Red Clay Project." This project concluded that erosion in the red clay area and subsequent transport and sedimentation is a natural process that has been intensified by forest harvesting, road building, and land-use conversions from forest to agriculture and open land (Andrews et al., 1980). Forest removal reduced evapotranspiration rates, resulting in a reduced resistance in soils to shear forces because of increased soil moisture content. In addition, the weaker root systems of replacement vegetation compared to the original white pine forests contribute to soil mass movement. The loss of large woody debris in stream channels also is thought to have reduced stream channel stability. In 1991, the conclusions of the Red Clay Project were reassessed, focusing attention on watershed factors that may be of greatest importance in correcting or mitigating the problem of soil mass wasting in the Nemadji basin. This chapter discusses the results of this reassessment.

SETTING

Soil mass wasting is the process of downslope movement of soil that consists of shear stress and displacement along surfaces that are either visible or that can reasonably be inferred (Huang,1983). Movement is thought to be the result of three contributing factors: (1) the high transportability rate potential of the soils; (2) the high soil moisture content; and (3) the absence of strong root systems to hold the soil in place. The Nemadji River Watershed exhibits excessive amounts of soil mass wasting (hereafter referred to as slumping). Most evident in the watershed are the chiefly rotational slumps that occur on hillslopes in the stream valleys and along the streams themselves. These slumps are suspected as a significant source of sediment to both the tributaries and the main stem of the Nemadji (Andrews et al.,1980).

The purpose of this research was to relate the frequency of slumping to watershed characteris-

Figure 11.1. Nemadji Basin location map.

tics that potentially have the greatest effect on soil mass wasting. These watershed characteristics were considered independent predictor variables of the dependent variable, frequency of slumping. Secondarily, the research was to determine data requirements for efficient and cost-effective development of a Geographic Information System (GIS) that could be applied for future watershed planning and management.

Based on previous work, this research analyzed the following nine watershed variables thought to affect soil mass wasting: stream channel gradient, time of concentration, stream length, degree of slope, total length of roads in the subwatershed, watershed area, percent coniferous cover, percent deciduous cover, and total forested area. Each of these variables was examined with frequency of slump sites in each of nine subwatersheds (Figure 11.2).

Fortunately, the relationships between forest cover and water yield have been the subject of considerable research. Results from research conducted in humid-temperate regions indicate that, in general, water yield is considerably greater under hardwoods than under conifers due to decreased interception losses and the longer dormant period of deciduous species (Swank and Douglas, 1974).

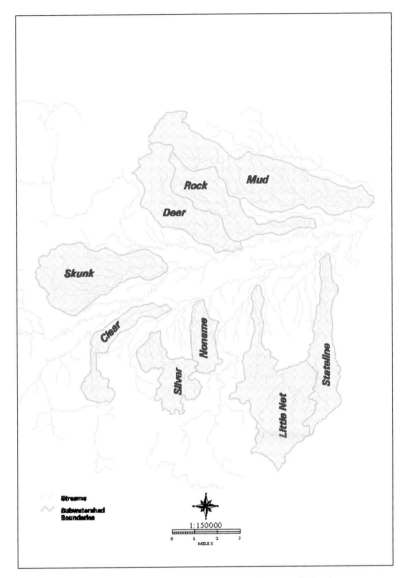

Figure 11.2. Map showing subwatersheds within the Nemadji Basin.

Increased streamflow provides additional energy for streambank degradation. Logging or thin-ning forests, converting from deep rooted to shallow rooted vegetation, or changing vegetative cover from one with a high interception capacity to one of lower capacity, all have been shown to increase water yield (Brooks et al., 1991).

Studies in Minnesota by Verry (1986) indicate that annual water yield increases and that aver-age annual peak flows can double after cutting upland forests. As the amount of nonforested area in a basin increases, one would expect water yield to increase and with it, higher streamflows with higher velocities and energy. Such conditions could promote the undercutting of slopes and conse-quently, streambank slumpage. Streambank undercutting, pervasive in the highly erosive red clays, leaves the banks susceptible to slides and slumping. Such effects could be greater following the removal of conifers than hardwoods.

In addition to the role of forest cover type on stream flow, the root strength of different species also can influence soil mass wasting (Sidle, 1985; Abe and Ziemer, 1991). The loss of tree roots diminishes resistance to slumping. On steep slopes bordering streams this combination induces soil creep and slumping. According to the Red Clay Project, high (prefrost) soil moisture levels can be positively correlated with the rate of spring soil mass soil wasting along stream banks (Andrews et al.,1980). Conifers, which transpire later into the fall and retain their interception capacity year-round, should reduce soil moisture during this critical period.

The dependent variable, slump frequency, was determined by tabulating areas where the mineral soil was exposed due to slumping; these sites were identified and mapped from high-resolution low-altitude aerial photographs. Frequency of slumps served as the primary indicator of the relative rate of erosion in each subbasin. In this application, slump frequency was evaluated for both entire subwatersheds and for a series of five discrete buffer zones surrounding each of the streams in those subwatersheds.

APPROACH

Pronounced differences in stream sediment concentrations and related turbidity exist among streams in the Nemadji Basin. To encompass the range of sediment-turbidity conditions, three subwatersheds were selected from each of three turbidity classes, low, moderate, and highly turbid. Classifications based on turbidity were verified using suspended sediment samples and discharge data collected during April and October 1993 and April 1994 (Table 11.1). Suspended sediment, turbidity, and streamflow discharge were measured for both upstream and downstream locations from each of the nine streams (USGS, 1977) at the locations shown in Figure 11.3 (see color section).

Each of the nine predictor variables and maps of slump sites were compiled for each of the nine subbasins in a vector-based GIS. A vector model was chosen for data development and analysis because these systems provide an excellent platform for development and mapping of point (e.g., slump sites), line (e.g., streams), and polygon (e.g., forest stands) data as well as for spatial data analysis. Additionally, local and regional management organizations had previously adopted vector GIS for in-house use. Design criteria for the Nemadji GIS data are map themes at a scale of 1:24,000, with Universal Transverse Mercator (UTM) Zone 15 coordinates applied to all spatial entities, built using conic projections. Base maps for the GIS were derived from existing USGS 7.5 minute quadrangle maps. Quad maps also served as source data for topographic (slope) data, stream locations, and stream gradient estimates. The generation of data themes, the crux of most GIS efforts, is described in the next two sections.

The three primary hydrologic themes are stream gradient, time of concentration, and watershed area. All blue-line (perennial and intermittent) streams were digitized from the USGS 1:24,000 quadrangle maps (Figure 11.4, see color section). The change in elevation from the headwaters to the mouth was taken directly from the quadrangle maps and divided by the length of the stream to determine gradient. Time of concentration was estimated using the Kirpich formula as reported by Gray (1970).

The primary terrestrial variables are slope, stream length, road length, and land cover; and the dependent variable, slump sites. Slope maps were interpolated from a series of contour lines digitized as points from the USGS quads, and a triangulated irregular network (TIN) algorithm was used to create an interpolated slope coverage for all basins. Mean slope was then calculated for each basin (Figure 11.4). The roads theme was acquired in digital vector format from the Minnesota Department of Transportation.

Stereo pairs of conventional color photographs, acquired in May 1992, were used to identify

Table 11.1. Summary of Basin Characteristics

High Turbidity   Moderate Turbidity   Low Turbidity

Variable	Gradient (m/km)	Discharge (L/s)[a]	S. Sediment (ppm)[a]	TOC (hr)	Stream Length (km)	No. Slumps	Mean Slope (%)	Road Length (km)	Total Area (ha)	% Conifer	% Deciduous	% Forested
Mud	5.3	17–382	6–35	5.37	2.22	45	52	49.2	3889	21	53	74
Rock	7.0	6–49	33–127	3.57	13.8	39	57	22.0	1830	22	33	55
Deer	7.9	54–106	43–105	3.25	11.1	45	50	13.5	2068	18	50	68
Skunk	5.7	26–371	6–30	3.95	14.3	27	66	33.3	2680	23	47	70
Noname	13.2	4–86	10–73	1.53	5.5	25	41	14.2	750	29	52	81
Clear	5.9	19–47	5–24	2.98	10.3	9	18	11.6	1031	29	44	73
Silver	9.1	98–193	1–17	2.13	6.9	2	36	11.7	1234	24	58	82
Little	7.8	66–357	1–11	4.71	19.5	13	52	18.7	3224	18	77	95
Net	12.3	36–95	4–12	3.38	14.3	25	49	11.3	1631	27	49	76
Stateline												

[a]Ranges are taken from three sampling dates (4/93, 10/93, and 4/94) and are reported for the downstream sampling point only.

and map slump sites. Photos were acquired from a helicopter platform using a 35 mm format camera. Negatives were printed at a nominal scale of 1:5,000, and interpreted using a 3-power mirror stereoscope. Land cover was mapped from stereo-pairs of 1:24,000 NHAP photographs into four land cover classes: (1) agricultural/open land; (2) coniferous forest; (3) deciduous forest; and (4) open water. NHAP prints were manually interpreted under magnification using a 0.5 hectare minimum mapping unit. Digital files of land cover were plotted and visually field-checked for accuracy of the cover typing (Figure 11.5, color section).

Slump sites were located on the 1:5,000 scale photos using a magnifying stereoscope. After intensive field checks (using a 75% sample) for accuracy and completeness, sites were entered into the GIS. Large-scale photos were required for accurate interpretation and each slump site could only be represented as a point location. However, due to differences in scale between the photos and the base map, it was difficult to locate the slumps precisely on the base map. Manual editing was employed to reposition the sites in some cases; as a result slump sites were estimated to be located to within 60 meters of their true location at a scale of 1:24,000 (Figure 11.6, color section).

The end result of data compilation was a 1:24,000 scale GIS for all of the nine subwatersheds. Values of land cover class area, ratio measures of percent cover, and lengths and values for the other variables as calculated within the GIS are reported in Table 11.1. The final data preprocessing step was to create data on buffer strips to examine the role of proximity of forest cover to stream channel as a covariate for the independent variables. To assess this effect, five buffer zones were defined along each stream channel, representing buffer strips of 10, 20, 40, 80, and 300 meters on each side of the stream.

Finally, a series of linear regressions were performed based on the frequency counts of soil mass movement sites in each basin and the measurements of the independent variables from the GIS.

RESULTS

The pattern of discharge follows the expected annual trend of highest flows occurring in April, with decreasing flows from August to October. The highest discharge gauged was just over 382 liters per second, measured in April on Skunk Creek. Although the low sample size prevents derivation of sediment rating curves and the use of statistical tests, there is no apparent difference in the observed discharges between the three classes that were grouped on the basis of stream turbidity. Overall, the more highly turbid streams (e.g., Mud, Rock, and Deer Creeks) not only had higher suspended sediment values, but they also exhibited more load variability over the sampling period.

Stream gradients are not higher in the more turbid streams (Table 11.1), and mean slope values are scarcely higher (but not statistically significant) for the more turbid streams. Total subwatershed areas vary from 3,889 hectares for Mud Creek to 750 hectares for Noname Creek; stream length and time of concentration are proportional to the total subwatershed areas for all of the basins. Mud, Rock, and Skunk Creeks have the highest values for total lengths of roads, and all are classified as having high to moderate turbidity. However, it is possible that road building may have its greatest effect at specific road cuts and bridge locations where slope stability was reduced.

Stream gradient was not significantly correlated to the number of slump sites and was relatively constant along the profiles. In a similar fashion, stream length did not exhibit a geographic pattern of variability, nor was it correlated to slump frequency. Time of concentration, conceptually, is an indication of the time it takes for water to travel from the hydraulically most distant point of a watershed to the mouth of the primary stream. The less the amount of time required, the faster a given stream will flow and thus the greater the energy potential for streamflow through the stream channel. No significant linear relationship was found between time of concentration and the number of observed slumps. However, there is little difference in time of concentration values among the different subwatersheds (Table 11.1).

The relative proportions of land-cover types varies considerably for the subwatersheds. The northern-most subwatersheds (Mud, Rock, and Deer Creeks) generally have lower forest cover area and higher suspended sediment concentrations (Table 11.1) than those subwatersheds to the south. Deciduous forest was the dominant forest type in all of the subwatersheds, with a maximum in the Little Net subwatershed of 77% and a minimum of 33% in Rock Creek basin. There was no strong geographic pattern of differences in the distribution of coniferous forest from one subwatershed to another.

As the width of the buffer zones increases, generally we see a reduction in the proportion of conifer cover to deciduous cover as a percentage of the area of the buffer. Across all nine subwatersheds it appeared that conifer cover types tended to occur most frequently in near-stream locations. When 300 meter widths were considered, the buffer strip usually extended to the ridgetop. This in part explains the increasing proportion of nonforested (agricultural and open) land at the 300 meter width.

The regression analysis of watershed characteristics and slump frequency yielded only one significant linear relationship. When the regressions were repeated for all of the buffer strip widths for all independent variables, again only one relationship was significant. In both cases forest cover percent was the significant variable. Note that the highest turbidity streams tend to have the most slump sites and a smaller relative proportion of forested land. Yet, low turbidity and high turbidity streams occur in watersheds with about the same percentage of conifer cover. The need to consider interactions of other variables and the distribution of the conifer cover in the basin is clear.

The percent slope of hillsides is generally assumed to be a significant covariate when examining causes of soil mass wasting. As percent slope increases, the effects of other contributing factors are amplified. These data show that as the mean slope of a subwatershed increases, there is a tendency for the number of slumps to increase as well. Although mean basin slope was not a significant predictor variable, it was noted that slope effects may be site specific, and when averaged over a large area, this effect was not apparent. Percent slope at each slump site, or the configuration of the hillslope associated with each slump site, could provide a different result.

The relationship between the total length of roads within each subwatershed and the number of slumps within that subwatershed was not significant. The larger the watershed, the greater the road length contained within it. Thus, the length of roads within a subwatershed was correlated with the size of the subwatersheds and not with the number of slumps. Results show the lack of a significant linear relationship of area to the dependent variables. As with slope percent, we suspect that roads need to be investigated on a site specific basis. The effect of roads most likely enhances slump occurrence at or very near the road cut itself. Because they were not mapped in this study, the impact of logging roads and skid trails remains untested.

On a subwatershed level, the occurrence of slumps generally did not diminish with increasing percent of conifer cover. There is a significant inverse relationship between the amount of total forest cover in a subwatershed and the number of slumps observed. As the amount of total forested area in a subwatershed increases the number of observed slumps decreases. However, when the buffer zones were analyzed this relationship did not persist. In fact, the only significant relationship found when repeating all regressions for all variables and all five buffer zone widths was the percent of coniferous cover within 80 meters of the stream channel. This might suggest that the percentage of the total watershed area in open/agricultural land affects the number of slumps, but the effect of location, or distribution, of the nonforested area in proximity to stream channels could not be tested at the scale and resolution of this study.

Study Implications

The GIS created for the basin contains maps of slump locations, land cover conditions, slope, and other characteristics for portions of the Nemadji River Watershed. In addition to serving as a data management tool, the GIS was used as the basis for data development and analysis. As a result of building GIS capability, several relationships between physical watershed characteristics could be examined systematically and efficiently.

Land cover was disproportionately distributed between the four land cover classes. Deciduous forest cover was more abundant in the upper reaches of the subwatersheds than in the lower reaches. In all subwatersheds, there is more deciduous cover than coniferous cover. Progressing downstream, the amount of conifer cover increased, especially within the stream valleys themselves. Much of the conifer cover was balsam fir (*Abies balsamea*) and spruce (*Picea* sp.) rather than the larger and deeper rooted white pine (*Pinus strobus*) that was the dominant conifer at the turn of the century. Conversely, the amount of open/agricultural area decreased from the upper to lower reaches of the stream. When large open areas occurred from the ridgetop leading down to a stream, there was generally a higher frequency of slumps in association with the open area.

Slope was identified as a significant covariate with slump locations from the Red Clay Project and other sources. Although our results did not show a significant linear correlation between percent slope and slope failure in the subwatersheds, there were plausible explanations. The mean slope and the number of slumps tended to increase closer to the mouth of the stream. Coincidentally, the percentage of conifer cover increased in lower areas as well. While the reason for the increase in conifer cover downstream was not known, it was possible that it would be more difficult to harvest on steep slopes.

Furthermore, the type of coniferous species present, which was not quantified from the aerial photographs, can influence slope stability through differences in root strength and rates of transpiration (for example, balsam fir is shallow-rooted and would not have the same slope stability effects of white pine or red pine). These two effects, among others, greatly complicate the analysis of cover type by slope class.

CUMULATIVE EFFECTS

The occurrence of soil mass wasting in the Nemadji Basin most likely was an expression of the cumulative effects of land use activities that have taken place over the past century. The removal of white pine forests at the turn of the century, the accompanying expansion of agricultural development, road construction, and the conversion from white pine to aspen-dominated forests all affected processes that would be expected to reduce slope and channel stability. The GIS approach allowed the consideration of individual variables as they might affect slumpage; however, the cumulative effects of all these activities over the watershed was not easily quantified and may be a situation of the total being "greater than the sum of the individual parts."

The removal of white pine and the subsequent replacement by either farmland or aspen forests would be expected to reduce slope strength because of loss of root strength on slopes. Overall reductions in evapotranspiration, leading to wetter soils, would increase soil stress. The loss of large woody debris in the channels would be expected to reduce channel stability, which coupled with increased water yield and stormflow volumes and peaks, would promote greater channel scour and streambank erosion. Road construction in the basin adds to these problems. Such processes occurring at the toe of the slope in these deeply incised channels would enhance soil slumps. After nearly 100 years, this watershed system still expresses, through excessive channel erosion, slumpage, and the resulting sediment loading, the cumulative effects of land use change. If we intend to develop models that help explain the cause-and-effect relationships of these systems, a better overall assessment of these cumulative effects is needed.

Because the percent of forested area was inversely related to a greater frequency of soil slumps, management should logically be aimed at increasing and maintaining forest cover in the watershed. Although these results could not identify preferred species for controlling slumps, the literature suggests that certain species such as white pine would be expected to provide more slope stability than other species such as aspen or birch. A more detailed look at species suitable for slope stability is warranted.

The results from this study underscore the need for future research into hillslope hydrology and fluvial processes in the Nemadji Watershed. This study points to certain factors that appear to be related to soil mass wasting and suspended sediment relationships in the Nemadji. Yet no definitive results exist to specifically address the actual relationships of water flow and subsequent soil mass wasting processes. More extensive streamflow data are needed in concert with hillslope process studies. Rainfall and snowmelt events over a range of magnitudes need to be studied so that we can better understand temporal relationships. In fact, the role that large rainfall events play in erosion is as yet unresearched in the Nemadji. Given these results and the availability of the GIS database, a process has been developed that enables managers to continue to examine watershed features and to monitor and track cumulative effects across this critical watershed.

REFERENCES

Abe, K., and R.R. Ziemer, 1991, Effect of tree roots on shallow-seated land slides. In *Proceedings, IUFRO Technical Session on Geomorphic Hazards in Managed Forests*. USDA Forest Service Gen. Tech. Rep. PNW–130, pp. 11–20.

Andrews, S.C., R.G. Christensen, and C.D. Wilson, 1980. Impact of Nonpoint Pollution Control on Western Lake Superior. Red Clay Project Final Report. U.S. Environmental Protection Agency Report, EPA 905/9-79-002-B.

Brooks, K., P. Folliott, H. Gregersen, and J. Thames, 1991. *Hydrology and the Management of Watersheds*. Iowa State University Press, Ames, IA.

Gray, D.M., Ed., 1970. Handbook on the Principles of Hydrology. Reprint. National Research Council of Canada. Port Washington: Water Information Center, Inc.

Huang, Y., 1983. *Stability Analysis of Earth Slopes*. Van Nostrand Reinhold Company, Inc., New York, NY.

Sidle, R.C., 1985. Factors influencing the stability of slopes. In *Proceedings, Workshop on Slope Stability: Problems and Solutions in Forest Management*. USDA Forest Service Gen. Tech. Rep. PNW-180, pp. 17–25.

Swank, W.T., and J.E. Douglas, 1974. Streamflow greatly reduced by converting deciduous hardwood stands to pine. *Science*, 185(4154):857–859.

United States Geological Survey, 1977. Sediment. In *The Handbook of Recommended Methods for Water Data Acquisition*. United States Geological Survey, Reston, VA, Chap. 3.

Verry, E.S. 1986. Forestry harvesting and water: The lake states experience. *Water Resources Bulletin*, 22(6):1039–1047.

CHAPTER 12

Development of a Database for Lake Ecosystem Studies: Linking GIS with RDBMS

Weihe Guan, Leslie J. Turner, and Sergio L. Lostal

INTRODUCTION

A comprehensive data management system is essential for any ecosystem study. In the Okee-chobee Systems Research Division, Ecosystem Restoration Department, South Florida Water Management District (SFWMD), large amounts of spatial data have been accumulated by years of Lake Okeechobee ecosystem studies. This chapter introduces the development of a database in support of the division's lake ecosystem studies. This database links a geographic information system (GIS) with a relational database management system (RDBMS) to best facilitate data use.

Developing and maintaining an integrated GIS-RDBMS database involves the following steps: (1) the establishment of a data server system conveniently accessible to all users; (2) the selection of appropriate RDBMS/GIS software for both attribute data and geographic data; (3) the development of a general data format and database structure; (4) the formalization of data management procedures, including input, update, conversion, QA/QC, and backup; and (5) the implementation of data query and retrieval utilities for end users to search, display, print, or plot information. This chapter documents the major steps in developing a Lake Okeechobee GIS-RDBMS database. The process focuses on the Lake Okeechobee ecosystem studies. Hardware and software availability impacted significantly the entire development approach. The chapter introduces a concept for modeling fuzzy geographic features, which addresses special needs in ecosystem studies. Also introduced in the chapter is the approach for establishing a virtual data server system through a computer network.

BACKGROUND

Lake Okeechobee (Figure 12.1) is the central feature of the Kissimmee River / Lake Okee-chobee / Everglades hydrologic ecosystem in south Florida. The lake is a large (approximately 700 square miles), shallow (average depth about 10 feet), subtropical lake which provides water, flood protection, and recreational benefits for a population exceeding 3.5 million people. The lake is also an important biological habitat for economically important fish and wildlife, including several threatened and endangered species (Aumen, 1995).

Various factors have contributed to the deterioration of the Lake Okeechobee ecosystem. Among these factors is excessive nutrient loading from agricultural activities in the watershed, which has caused increased blue-green algal blooms. These blooms, characterized by surface

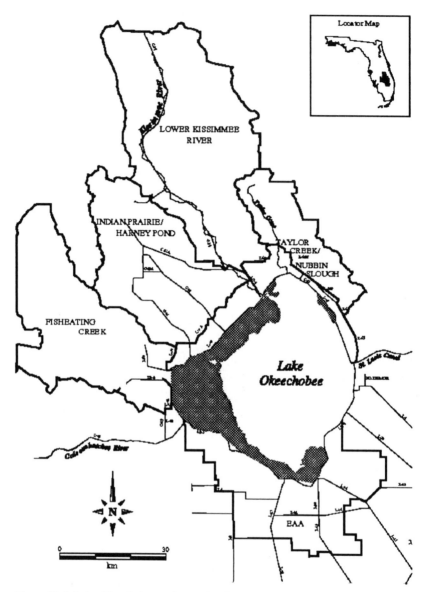

Figure 12.1. Lake Okeechobee and associated ecosystem studies.

scums and unpleasant tastes and odors, raised concerns about declining water quality (Aumen, 1995). Several interdisciplinary, multiyear research efforts were initiated in the late 1980s in response to the algal blooms, including a lake ecosystem study.

The Lake Okeechobee Ecosystem Study (LOES), conducted by the University of Florida under a contract with the South Florida Water Management District (SFWMD), was unique in that it looked beyond excessive nutrient loading to other components of the ecosystem. Research topics include water level effects, water quality, fish and wading bird populations, and wildlife habitat. The study's objective was to provide an ecological baseline against which future ecosystem trends can be compared, and to assess the general health of the ecosystem. The database structure that will be described in this chapter is based on the results of this LOES.

Data Related to Lake Okeechobee Ecosystem Study

The Lake Okeechobee Ecosystem Study addressed the following issues (University of Florida, 1991): (1) data synthesis, modeling, and database management; (2) water chemistry and physical parameters; (3) community and ecosystem ecology of emergent macrophytes; (4) phytoplankton, bacteria, epiphytes, submerged plants, macroinvertebrates, and zooplankton; (5) distribution and abundance patterns and the reproductivity and foraging ecology of wading birds; and (6) larval and juvenile fish. The project involved extensive field data collection and analysis. Data related to the study were archived on floppy disks using Lotus 1–2–3 spreadsheets, WordPerfect documents, ASCII text files, and ERDAS raster files. Data were categorized as follows: (1) plankton—bacteria, bioassays, nitrogen fixation, phytoplankton, and zooplankton; (2) plants—emergent, nutrients, seeds, soils, and submergent; (3) water quality—chlorophyll, nutrients, physical chemistry, and suspended solids; (4) wildlife—birds and fish; (5) hydrology—Lake Okeechobee hydrological data; (6) spatial data—GIS coverages, images, and locational files; and (7) documentation—various text files for clarification, identification, and explanation.

Importance of GIS

Most of the data collected in LOES have locational records. Location is recorded either by x-y coordinates or by verbal descriptions. The geographic information system (GIS) was identified as an important tool for the lake ecosystem study. As stated in a LOES annual report (University of Florida, 1991), the objective of LOES was to use an ecosystems approach to develop a set of tools (models and GIS databases) that integrate the data from various tasks in the project and other projects to provide predictive capabilities with which the SFWMD can evaluate the consequences of various water management options on the marsh littoral zone of Lake Okeechobee and its interaction with the pelagic zone of the lake. Critical concerns include impacts on the fish and wildlife resources, the role of the marsh in nutrient dynamics and exchange with the lake, and estimates of the total flux of nutrients to and from the lake under different water regimes. The following areas of focus were suggested by the LOES review panel (University of Florida, 1991): (1) impact of lake stage on biotic communities; (2) impact of nutrient concentrations on biotic communities and trophic relationships; (3) direct and indirect effects of plant community structure on critical habitat and energy flow to wading birds and fishes; (4) role of littoral zone in the lake's ecology; (5) effects of water pumping from canals on lake communities and productivity; and (6) role of exotics in lake ecology. These focus areas outlined user database requirements.

The LOES review panel (University of Florida, 1991) suggested that a spatially based predictive model utilizing a GIS approach be developed, capable of predicting the responses of fish and wildlife resources to management options such as lake stage manipulation, nutrient loading increases or decreases, or the long-term effects of maintaining the present regimes of stage, nutrients and flows. The model was envisioned to be sufficient to provide indications of the magnitude of the changes in the system and the spatial location of such changes utilizing the GIS information layers and model parameters derived from the various tasks of LOES and other projects.

To effectively use information collected by LOES and other studies, a functional database linking GIS with a relational database management system (RDBMS) was needed. An RDBMS can efficiently manage the large amount of information while a GIS can present data spatially. The effort of developing this database included establishing a data server system, designing an integrated database structure, developing a database management procedure, and implementing a data query and retrieval user interface.

DATA SERVER SYSTEM

A data server system is a computing platform that hosts databases. It may include computers, operating systems, networks, database management systems, geographic information systems, and other hardware and software necessary to input, store, manage, query, and output data (Cowen et al., 1995). Ideally, a data server system should be selected according to the conceptual design of the database to be developed. In reality, many constraints, especially financial, limit available options for the selection of a data server system.

In this study, the computing environment consists of ORACLE as the RDBMS software on a VAX minicomputer, and ARC/INFO and Arcview as the GIS software on SUN SPARC workstations. Some workstations use SUN OS and others use Solaris as the operating system. All computers were networked. Desktop workstations (SPARC 2, 10, 20, or Ultra) and personal computers were available to end users of the GIS-RDBMS database.

The GIS database was a special component of a relational database hosting the information collected by LOES. The GIS database had two components: ARC/INFO coverages for geographic features and unique feature IDs, and ORACLE tables for attribute items with the feature ID as a unique key. The connection between the two was built on the Database Integrator in ARC/INFO and ArcView.

Due to software and storage space limitations, the database cannot be loaded onto a single computer. Several networked computers were needed. Tabular data are stored in ORACLE tables on the VAX, and spatial data were stored in ARC/INFO coverage format on several workstations. One workstation, a SUN SPARC 20, was designated as the "virtual" data server. The graphic user interface for data query and retrieval was installed on this workstation, with the directory structure for all ARC/INFO coverages. Symbolic links in subdirectories point to the actual locations of the coverages on other workstations. When users query on spatial features from a workstation, an ARC-ORACLE interface links to the ORACLE database on VAX and returns appropriate tables and records (Figure 12.2).

This design provides users with a seemingly holistic database on the virtual data server. Users need not know the real storage locations of the database components, nor do they need to interact with any computer platform other than the virtual data server. On the other hand, the group of networked computers collectively provides the required storage and computing capacity necessary for the database, which does not exist in any single computer. When the database grows, more computers may be brought into the group with minimum impact on existing server components.

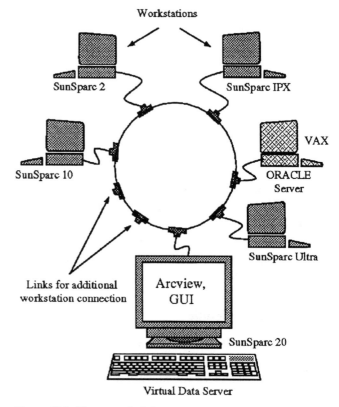

Figure 12.2. The networked data server system.

The disadvantage of this data server structure is its reliance on the network and each computer in the structure. If one data-hosting computer goes off the network, the database will not function properly. Moreover, the database manager must have access to all database computers for maintenance purposes. Because most of the involved workstations are routinely used by SFWMD staff as desktop computers, special effort is needed to coordinate their use for the database.

THE DATABASE MODEL

A model can be thought of as a real-world abstraction where only essential details are kept. Database models are created to understand the data organization before the database and support programs are built. The LOES database model was created using the Object-Oriented Modeling Technique (OMT) (Rumbaugh et al., 1991). Subsequently, the LOES database model was mapped as a relational database and built on a VAX minicomputer using an ORACLE RDBMS.

Originally, LOES data were organized in spreadsheets with different record formats. After data collection was completed, scientists developed a database model from which data could be extracted in convenient formats. The resulting database model has five modules (Lostal, 1996): (1) Locational, (2) Ecological Variable, (3) Biological Species, (4) Time Series, and (5) Organizational.

Locational, Ecological Variable, Biological Species, and Organizational modules have separate, well-defined purposes. The Locational module deals with the measurement site or station. The Ecological Variable module handles information about the type of data (i.e., sample method, parameter, and unit) measured. The Biological Species module describes scientific and common names of measured species. The Organizational module addresses who (i.e., agency, observer) produced the data and why (i.e., project, study, task) the data were produced.

Time Series, the last module, depends on the other modules. Time series are determined by stations, ecological variables, and biological species; they contain summary information and a data point set. Summary information includes items such as period of record, number of observations, average or sum, standard deviation, and maximum and minimum values. The data point set is formed by all data points observed. Each data point encompasses a time series identifier, time stamp, measured or described ecological value, quality indicator information, and organizational code.

The database's architecture allows spatial queries about ecological data. For example, whenever a user selects a location in a map coverage, the GIS client program sends the request to a utility program that chooses the station closest to that location. The program queries the database using the selected station and the database returns a list with all the time series summaries stored for that station. The user reviews the list, selects an appropriate time series, and submits the request. The program queries the database, receives the information, and displays the data point set. As the example shows, the ORACLE database and the GIS client interface initially through locational information.

The Locational Module

The LOES database was modeled using object-oriented methods. As its name suggests, the object-oriented approach organizes real-world concepts as objects. Objects are things or abstractions with boundaries and meaning in the real world. Objects with similar properties are grouped into classes. A database model shows the classes that compose a database, and how these classes associate with each other. The Locational module deals with the classes that represent the measure-

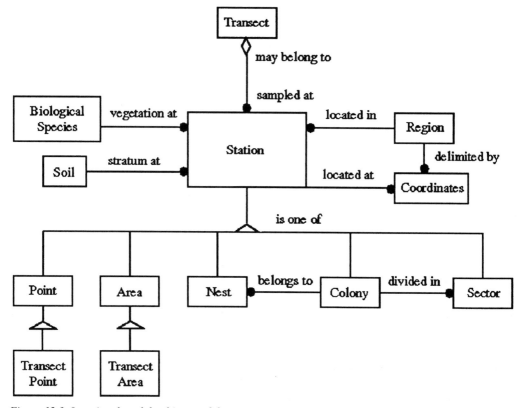

Figure 12.3. Locational module object model.

ment site and their associations. The GIS client is interested particularly in this module, because any information displayed on a map coverage is attached to a measurement site.

Figure 12.3 shows the database model for the Locational module. The measurement site, depicted by the Station class (classes are represented as rectangles in the diagram) is the fundamental component of this module. All classes included in this module associate with the Station class (associations are represented by lines in the diagram). Some classes are specific location types, other classes associate spatially with the Station class, and the rest provide complementary information to the Station class.

Ecological data were collected at a variety of places such as water quality monitoring points, fishing sites, and bird colonies. The Station class was an abstraction that represents any location type. This special association, called a generalization (represented by a triangle in the diagram), was used to classify related classes. As the figure shows, the Station class was a generalization of five location types used to collect data: (1) Point, (2) Area, (3) Bird Colony, (4) Colony Sector, and (5) Bird Nest. Points are sites identified by one pair of x-y coordinates. Areas refer to locations specified by more than one pair of x-y coordinates (two points define a rectangle). Other station types describe bird information. Bird colonies were sites where birds live and procreate. Both colony and point positions were given by one pair of x-y coordinates. However, colonies were classified separately from points because they also associate with other station types such as nests and colony sectors. The use of labels helps to interpret associations. For example, the association between the Nest class and the Colony class indicates that many nests (the filled circle means many) belong to a colony.

Spatially, the Station class associates with regions and transects. Regions are large areas where stations were located. The diagram indicates that many stations were located in a region. Transects are station aggregates sampled along a transect path. This is a "whole-part" association or aggregation represented as a diamond in the diagram. The transect is the total assembly and the stations are its elements. Of the five station types, only point and area stations may belong to transects. This aspect is shown on the figure by representing Transect Point and Transect Area as separated classes.

The Coordinates classes were related to both Region and Station classes. The Region-Coordinates association shows that a region was delimited (specified) by many coordinates. The association between the Station class and the Coordinates class was more general. The figure indicates solely that a station is located at many coordinates. This association considers all location types explained above. First, a station refers to a location type indicated by one set of x-y coordinates. Second, the station was an area determined by more than one set of x-y coordinates. Third, a station position may be reported without using coordinates values; instead, a descriptive method was used (See Soft Points and Soft Polygons section).

The last two classes in the figure, Biological Species and Soil, serve as "catalogs" for the Station class. For example, the Biological Species class contains all information about vegetation including scientific classification and common name. However, instead of duplicating all the information, the Station class contains solely abbreviated descriptions. When a full description is required, the Biological Species and the Soil classes supply the information that is not stored in the Station class.

Soft Points and Soft Polygons

Due to the biological and ecological nature of the data sets, the LOES database developers needed to process spatial data with uncertain locational coordinates. To address this issue, we introduced the concept of soft points and soft polygons. A soft point is a point without a definite x-y location, and a soft polygon is a polygon without a definite boundary. In ecosystem studies, researchers often have to deal with soft points and polygons when historical and field survey data are involved. Before GIS was implemented, many field observations were made with a verbal description of location, not explicit x-y coordinates. Even with well-established GIS concepts, some ecological features were difficult to describe at a definite x-y location or within a distinct boundary. For example, a given fish species was observed in a certain area of a water body at a certain time. That "certain area" of the water body does not have a clear boundary. Such observations may apply to animals on land, plankton in water, or a floating plant mass in a wetland.

In the soft feature model, the geographic location of a soft point was described by its probability distribution in space. A soft point may appear at any known location with a certain probability. That known location was usually contained in a polygon. When the probability equals one at a known point, the soft point becomes a hard point. Where the probability equals zero, the soft point never appears. The line between none-zero and zero probability areas was the boundary of the probability distribution zone. One soft point may have multiple probability distribution polygons (Figure 12.4).

The geographic location of a soft polygon was more difficult to describe than that of a soft point. Three parameters were required to define a polygon: size, shape, and the location of the gravity center. When any of these parameters is uncertain, the polygon becomes a soft polygon. In theory, this leads to seven types of soft polygons (Table 12.1). In this study, two types of soft polygons are discussed (Types 1 and 5 in Table 12.1), which are most common to ecological studies.

For soft polygons with uncertain size, shape and location, the probability distribution patterns were similar to those of soft points. The probability polygons were stored as an ARC/INFO cover-

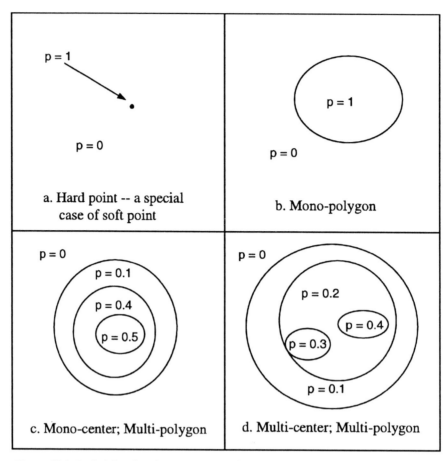

Figure 12.4. Probability distribution patterns of soft points.

age, each with an attribute value indicating the probability of any point in the probability polygon belonging to the soft polygon.

For soft polygons with known size and shape, the only uncertainty was location. The polygon's gravity center may be derived from its size and shape. The probability distribution of the center determines the probability distribution of the polygon. The soft point model discussed above also applies to the center of this soft polygon type. The size and shape of the polygon can be preserved

Table 12.1. Types of Soft Polygons, sort by size, then by shape, then by center location

Type	Size	Shape	Center Location	Note
1	uncertain	uncertain	uncertain	common
2	certain	uncertain	uncertain	uncommon
3	uncertain	certain	uncertain	rare
4	uncertain	uncertain	certain	uncommon
5	certain	certain	uncertain	common
6	certain	uncertain	certain	uncommon
7	uncertain	certain	certain	rare
8	certain	certain	certain	hard polygon, a special case of soft polygon

in a separate coverage or incorporated into the polygon coverage for probability distribution through an appropriate algorithm.

DATABASE IMPLEMENTATION AND MANAGEMENT

Initial database implementation included the following: (1) specifying a directory structure; (2) determining a directory and file-naming convention; (3) specifying read/write permissions and work group arrangements; (4) selecting precision and projection systems for coverages; (5) setting up a standard template for attribute files; and (6) establishing metadata standards.

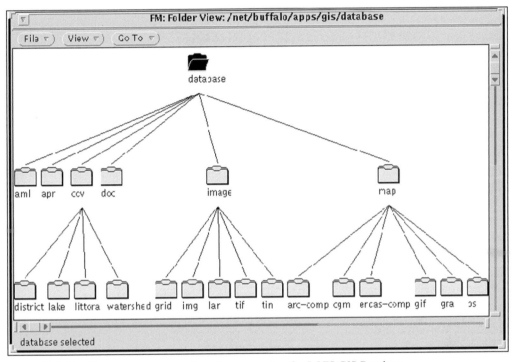

Figure 12.5. The directory structure and naming convention for LOES GIS Database.

The GIS database directory structure was set up on the virtual server. It includes subdirectories for ARC/INFO coverages, ArcView project files, images, map files, programming codes, and documentation files. A naming convention was developed to indicate the subject and format of each subdirectory and file (Figure 12.5).

Access permissions were assigned according to whether an individual was a database developer or user. Developers include database managers, system administrators, and data editors. Database managers have full read/write access to the database and are responsible for managing both the GIS database and the data server system. The system administrators, who also have full read/write access, solve system problems, maintain system security, and perform periodic backups by saving data on external media. Data editors have partial write access to input, update, and/or convert data for the database as well as utilizing metadata standards to document information about the database. End users are an integral part of GIS database management as they provide valuable feedback for database improvement. All end users have read-only access to the database.

Single precision (seven significant digits) and double precision (15 significant digits) are the only alternatives for storing coverage coordinates in ARC/INFO. Double precision coverages, which provide a more precise geographic location, require more storage space. Due to the large study area (hundreds of square miles) and the variation of parameters within a short distance (inches or feet) in the Lake Okeechobee ecosystem, double precision was used for most coverages.

The projection system was based on the current GIS standard of the SFWMD, which is State Plane zone 3601 (Florida East Zone). The datum for the database was NAD27 (1927 North American Datum). When the SFWMD GIS database migrates to NAD83 (1983 North American Datum), all coverages and images in the Lake Okeechobee database will be transformed accordingly. The transformation will, most likely, use the ARC/INFO PROJECT function.

Minimal attribute data were stored with the coverages in the GIS database. Most ecological data reside in ORACLE tables and are linked to geographic features in the coverages by a unique ID. This ID often is the only external (user specified) attribute in the coverages. The user-specified attribute item (UNIQUE-ID) is indexed to decrease process time across the Arc-Oracle link.

In order to standardize metadata format for the GIS databases, METAMENU, a menu-driven metadata editing user interface was developed. METAMENU provides a convenient tool for users to document GIS data following a predefined standard. It automatically extracts any existing metadata information from GIS files, provides multiple choices whenever applicable, and prompts users to enter required information. The interface also checks for completion of user input, reorganizes entries into a standard metadata format, and saves information at a logical location using a standard naming convention. With this interface, users may define one metadata format for a group of similar coverages, or copy the metadata of one coverage into another and selectively edit some items to document differences.

One major difference between METAMENU and the DOCUMENT command in Arc/Info ver.7 is the metadata file format. DOCUMENT saves metadata in INFO, while the METAMENU interface saves metadata as an ASCII text file. ASCII files can be viewed without an Arc/Info license. Moreover, METAMENU generates metadata more specific to the LOES database users' needs, while DOCUMENT is more general and was designed for a broad range of ARC/INFO users. An example of metadata generated using METAMENU is in Appendix 12.1.

Database Management

The management of a GIS database, similar to that of other databases, involves data development, QA/QC, and backup. Data development includes data input, update, conversion, and construction of metadata. QA/QC, which stands for quality assurance and quality control, is a process following data development. Backup safeguards data from damage.

The unique aspect of GIS data management is the coordination between geographic and attribute data. Geographic data require special techniques and procedures for input, update, conversion, QA/QC, and backup; corresponding attribute data can be managed as regular tabular data. The geographic features and tabular attributes are linked together through a unique ID, which requires special attention when either side of the database is modified.

Data development, QA/QC, and backup of GIS data become more difficult to manage in a multiuser environment. A GIS data management guideline was a necessary tool which assists both database managers and users in systematically handling such transactions. The Lake Okeechobee database management utilized the GIS Data Management Guidelines. These guidelines detail the responsibilities of the database managers, the system administrators, the data editors, and the end users. They also specify the procedures for data QA/QC (SFWMD, 1997).

After data development was completed, data were verified by manual QA/QC procedures. Criteria for assessing data quality were specified for both geographic features and their attributes. According to Montgomery and Schuch (1993), graphic and attribute data "have similar data quality components, such as completeness, correctness, timeliness, and coverage, but the graphic features also are subject to cartographic quality considerations . . . (such as) relative accuracy, absolute accuracy, and graphic quality." The QA/QC procedures for Lake Okeechobee database incorporated these considerations in the data verification process.

An integral part of database management includes database backup. Backup saves database contents on external media as a safeguard against corruption. Three kinds of backup procedures were outlined in the management guidelines for the Lake Okeechobee database: (1) routine backups, which are consistently done at a designated time, regardless of changes in data; (2) nonemergency backups, which are done after changes in data; and (3) emergency backups, which take place preceding the possibility of a natural disaster (e.g., hurricane), power failure, maintenance on critical hard/software, and as otherwise deemed necessary.

DATA QUERY AND RETRIEVAL

The GIS database is accessible to all SFWMD staff for data query and retrieval. A generic user interface is being developed to support users with minimum GIS training to search, display, print or plot data from the database. Thus, the interface must be tailored to user needs.

The first step in interface development was to interview end users and identify their needs for data display, query, and retrieval (Montgomery and Schuch, 1993). Considering the internal structure of the database, user demand, and facilities available, ArcView was selected as the interface platform. Avenue, the object-oriented scripting language for ArcView, was used to communicate with the ORACLE database, customize the display environment, structure query statements, and standardize output formats. A prototype of the interface has been developed, presenting the interface look-and-feel. Development was based on user comments. The interface will be modified based on further input from users.

SUMMARY

This chapter documents the major steps in developing a Lake Okeechobee GIS-RDBMS database. The entire process involved the following components: (1) data server setup; (2) data model design; (3) database implementation; (4) data management standardization; and (5) user interface development. Database development focused on the Lake Okeechobee Ecosystem Study. The approaches adopted in database development were largely determined by hardware and software availability as well as special characteristics of ecological data. Introduced in the chapter is a method for modeling soft geographic features. The chapter also describes the approach for establishing a virtual data server through the network.

Database development is a constant trade-off between "what it should be" and "what it could be." Instead of a "perfect database," the final product of this project is a "functional database," developed under various constraints. Database optimization is a long-term task. By having a functional database first, and improving it constantly, the database will incrementally approach an ideal design that best serves its users.

ACKNOWLEDGMENT

This chapter incorporated review comments from Todd Tisdale, Al Steinman, Chris Carlson, Benjamin Lewis, and Nancy Lin. The metadata editing user interface used a set of AML scripts from a command-drive metadata input program originally developed by Timothy Liebermann. The data query and retrieval graphic user interface was developed on an ArcView 2.0 data viewing project initially built by Chris Carlson. The authors wish to thank the above-mentioned colleagues—all are staff of the South Florida Water Management District—for their valuable contributions to this study.

REFERENCES

Aumen, N.G., 1995. The history of human impacts, lake management, and limnological research on Lake Okeechobee, Florida (USA). In *Advances in Limnology, Ecological Studies of the Littoral and Pelagic Systems of Lake Okeechobee, Florida*. N.G. Aumen and R.G. Wetzel, Eds., Schweizerbart, Stuttgart, Germany, Vol.45, pp.1–16.

Cowen, D.J., J.R. Jenson, P.J. Bresnahan, G.B. Ehler, D. Graves, X. Huang, C. Wiesner, and H.E. Mackey Jr., 1995. The design and implementation of an integrated geographic information system for environmental applications. *Photogrammetric Engineering and Remote Sensing*, 61(11):1393–1404.

Lostal, S.L., 1996. Software Development for Ecological Data System. Florida Atlantic University, Boca Raton, FL.

Montgomery, G.E., and H.C. Schuch, 1993. *GIS Data Conversion Handbook*, GIS World, Inc., Fort Collins, CO.

Rumbaugh, J. et al. 1991. *Object-Oriented Modeling and Design*, Prentice Hall, Englewood Cliffs, NJ.

SFWMD, 1997. The GIS Data Management Guidelines for Lake Okeechobee Database, unpublished.

University of Florida, 1991. Ecological Studies of the Littoral and Pelagic System of Lake Okeechobee. Annual report prepared for South Florida Water Management District, West Palm Beach, FL.

Appendix 12.1. An Example of Metadata Generated Using METAMENU

```
H
H       SOUTH FLORIDA WATER MANAGEMENT DISTRICT
H            3301 Gun Club Road
H            PO Box 24680
H            West Palm Beach, FL, USA 33418–4680
H            (800) 432–2045
H            (561) 686–8800
H
H       GEOGRAPHIC INFORMATION SYSTEMS METADATA
H
H
H       DOCUMENTATION HISTORY
H       ArcMeta Rev. 2.0, lmoore, 10/30/97.11:22:54.Thu.
H            Metafile Name: lo_dairy97.met
H            Cover: lo_dairy97, Workspace: /home/kos/lmoore/gis/dairy_covs
H            Type: cover, Action: template
D
D       ARC/INFO DESCRIPTION of cover geodataset LO_DAIRY97
D            (Created by lmoore on 10/30/97.11:22:57.Thu)
D            Workspace: /home/kos/lmoore/gis/dairy_covs
D            Geodataset: lo_dairy97
D
D       COORDINATE INFORMATION:
D            Precision: DOUBLE
D            Minimum X and Y values: 472335.2808281, 1028100.9375
D            Maximum X and Y values: 620659.125, 1137946.719614
D            Latitude/Longitude Minimum Bound: 27.1413, –81.1082
D            Latitude/Longitude Maximum Bound: 27.4848, –80.605
D
D       PROJECTION INFORMATION:
D            Projection Name: STATEPLANE
D            Units: FEET
D            Zone: 3601
D            Datum: NAD27
D            Spheroid: CLARKE1866
D            False Easting: 0
D            False Northing: 0
D
D       COVERAGE INFORMATION:
D            Coverage Pathname: /HOME/KOS/LMOORE/GIS/DAIRY_COVS/LO_DAIRY97
D            Number of Polygons: 3036
D            Number of Arcs: 9877
D            Number of Segments: 59870
D            Number of Annotations: 0
D            Number of Tics: 22
D            Fuzzy Tolerance: 0.5
D            Dangle Distance: 1
D
D       POLYGON INFORMATION:
D            Number of Polygons: 3036
D            Bytes in PAT: 66
D            Polygons Indexed: .FALSE.
```

Appendix 12.1. (Continued)

```
N
N       NARRATIVE DESCRIPTION of cover geodataset LO_DAIRY97
N             (Created by lmoore on 10/30/97.11:23:25.Thu)
N
N       GENERAL INFORMATION
N             Short Description: A polygon coverage of active dairies in the LO watershed and their associ-
              ated landuses.
N             Geographic Extent: Okeechobee county
N             Accuracy: same as /net/b50home1/proj/erd/gis/kos2/dairies/covs/dairy91
N             Abstract 1: Updated with current information from dairy landowners and Okeechobee Service
              Center personnel.
N             Abstract 2:
N             Abstract 3:
N             Abstract 4:
N             Intended Use 1: LOADSS model runs
N             Intended Use 2:
N             Intended Use 3:
N             Intended Use 4:
N             Limitations 1: same as /net/b50home1/proj/erd/gis/kos2/dairies/covs/dairy91
N             Limitations 2:
N             Limitations 3:
N             Limitations 4:
N
N       DISTRIBUTION INFORMATION
N             Availability: internal
N             Distribution Format: coverage through network
N             Distribution Size (Mb): 1157
N
N       CONTACT INFORMATION
N             Contact Person: Leslie J. Turner or Weihe Guan
N             Contact E-Mail Address: wguan@sfwmd.gov
N             Contact Phone Number: (561)687–6610 or (561)687–6687
N             Contact Address: OSRD, ERD, SFWMD
N
N       SOURCE INFORMATION
N             Source Description: same as /net/b50home1/proj/erd/gis/kos2/dairies/covs/dairy91
N             Source Media:
N             Source Scale:
N             Source Date: 1993
N             Source Contact Person: Joyce Zhang or Weihe Guan
N             Source E-Mail Address: jzhang @sfwmd.gov or wguan@sfwmd.gov
N             Source Phone Number: (561)687–6341 or (561)687–6687
N             Source Address: OSRD, ERD, SFWMD
N
N       ADDITIONAL INFORMATION
N             Related Information 1: None
N             Related Information 2:
N             Related Information 3:
N             Related Information 4:
N             Attribute Description 1:
N             Attribute Description 2:
N             Attribute Description 3:
N             Attribute Description 4:
```

Appendix 12.1. (Continued)

```
N          Processing History 1:
N          Processing History 2:
N          Processing History 3:
N          Processing History 4:
N          Revisions 1:
N          Revisions 2:
N          Revisions 3:
N          Revisions 4:
N          Misc. Notes 1:
N          Misc. Notes 2:
N          Misc. Notes 3:
N          Misc. Notes 4:
|
|          INFO ITEM DEFINITIONS for poly LO_DAIRY97
|          (Entered by lmoore on 10/30/97.11:23:31.Thu)
|
|          ITEM NAME: AREA
|               Item INFO Definition: 8,18,F,5
|
|          ITEM NAME: PERIMETER
|               Item INFO Definition: 8,18,F,5
|
|          ITEM NAME: LO_DAIRY97#
|               Item INFO Definition: 4,5,B,0
|
|          ITEM NAME: LO_DAIRY97-ID
|               Item INFO Definition: 4,5,B,0
|               Short Description:
|               Valid Codes:
|               Data Source:
|               Accuracy:
|               Related INFO Table:
|               Narrative 1:
|               Narrative 2:
|
|          ITEM NAME: TAG
|               Item INFO Definition: 4,4,C,0
|               Short Description: Landuse codes.
|               Valid Codes: IMPA, BARN, OTFL, HIA, MHP, DOL, OTP, SSA, SPFL, WSP, WET, CSF
|               Data Source: same as /net/b50home1/proj/erd/gis/kos2/dairies/covs/dairy91
|               Accuracy:
|               Related INFO Table:
|               Narrative 1:
|               Narrative 2:
|
|          ITEM NAME: DNAME
|               Item INFO Definition: 12,12,C,0
|               Short Description: Dairy names.
|               Valid Codes:
|               Data Source: same as /net/b50home1/proj/erd/gis/kos2/dairies/covs/dairy91
|               Accuracy: same as /net/b50home1/proj/erd/gis/kos2/dairies/covs/dairy91
|               Related INFO Table:
|               Narrative 1:
```

Appendix 12.1. (Continued)

| Narrative 2:
|
| ITEM NAME: LANDUSE
| Item INFO Definition: 25,25,C,0
| Short Description:
| Valid Codes:
| Data Source:
| Accuracy: same as /net/b50home1/proj/erd/gis/kos2/dairies/covs/dairy91
| Related INFO Table:
| Narrative 1:
| Narrative 2:

CHAPTER 13

Historical Aerial Photographs and a Geographic Information System (GIS) to Determine Effects of Long-Term Water Level Fluctuations on Wetlands along the St. Marys River, Michigan, USA

Donald C. Williams and John G. Lyon

INTRODUCTION

Water levels of the Laurentian Great Lakes and their connecting channels rise and fall in predictable annual patterns related to the seasonal patterns of precipitation, runoff, and evaporation. Variations of winter low levels to summer high levels average from 0.37 meters on Lake Superior to 0.6 meters on Lake Ontario (U.S. Army Corps of Engineers, 1985). Lake levels also rise and decline in unpredictable multiyear patterns following long-term fluctuations of climatic conditions in and around the Great Lakes basin. Historically, the ranges of Great Lake annual water level averages have been from ± 0.6 to 0.9 meters from their respective historic means (U.S. Army Corps of Engineers, 1985).

The Great Lakes long-term water level fluctuations have recurred about every two to three decades. These fluctuations have broad influences on the areas of the wetlands situated along the lake and connecting channel shores (Jaworski et al., 1979; Harris et al., 1981; Enslin and McIntosh, 1982; Lyon and Drobney, 1984; Busch and Lewis, 1984; Quinlan, 1985; Greene 1987; Bukata et al., 1988; Williams, 1995). The subject of this chapter is the use of a Geographic Information System (GIS) to study the influences of long-term water level fluctuations on the wetland areas of the St. Marys River. The St. Marys River is a connecting channel between Lakes Superior and Huron that borders on the U.S. and Canada. The river water levels are largely controlled by the levels of Lake Huron (Williams, 1995), such that wetland areas along the St. Marys undergo fluctuations corresponding to those of Great Lakes wetlands. A GIS was used to measure area changes in five different St. Marys River coastal wetland classes. These area change data were used to characterize wetland class responses to water level changes, to estimate response times, and to examine how wetland class transfers were affected by water levels.

Geographic Information Systems (GIS) offer excellent capabilities as tools for analysis of wetland changes caused by long-term lake level fluctuations. When accurate historical information on wetland areas and wetland characteristics is available in digital form, it would appear that GIS change detection algorithms could be employed to rapidly provide information on changes in wet-

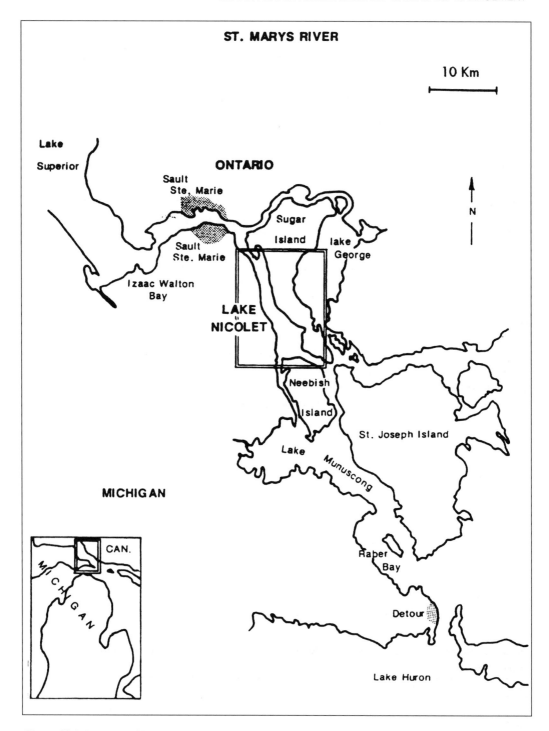

Figure 13.1. Location of St. Marys River study area.

land area and changes in wetland vegetation class associated with water level fluctuations. The following analysis demonstrates the application of such methods to analyze wetland changes due to long-term Great Lakes water level fluctuations. This analysis was conducted on a U.S. Fish and Wildlife Service (USFWS) National Wetland Inventory digital data set on the wetlands of the St. Marys River. This analysis used ERDAS GIS software. Other GIS systems containing similar analytic software would provide similar results.

The St. Marys River connects Whitefish Bay, Lake Superior, with Detour Passage in northern Lake Huron. The wetland study area was directly south of Sault Ste. Marie, Michigan, from Little Rapids Cut at the northern end of Sugar Island, south along both sides of Lake Nicolet, and along the east side of Neebish Island into northern Lake Munuscong (Figure 13.1). The study area is covered by five United States Geological Survey (USGS) 1:24,000 (7 1/2 minute) quadrangle maps.

The wetland types found in the study area are characterized as Unconsolidated Bottom, Emergent Wetland, Unconsolidated Shore, Scrub-Shrub Wetland, and Forested Wetland, using the USFWS Classification System (Cowardin et al., 1979). These wetlands classes are commonly described in nontechnical terms as river bottom, marsh, beach, scrub-shrub, and swamp.

METHODS

Wetland classes and areas in the St. Marys River were determined by interpretation of aerial photos from the summer seasons of 1939, 1953, 1964, 1978, 1982, 1984, and 1985 (Table 13.1). The film types included black and white, black and white infrared, color, and color infrared. The photo scales ranged from 1:12,000 to 1:58,000 (Table 13.1).

The aerial photos were acquired at seven different St. Marys River water levels (Table 13.1). These levels represented a broad sample of the wide range of water levels to which Great Lakes coastal wetlands are subjected. The August 1953 level, 177.52 m, was within 20 cm of the all-time monthly high of 177.64 m August '52 at the U.S. Slip gauge. The July 1964 level, 176.09, was within 25 cm of the all-time low of 175.86 m (Dec. '63). These aerial photos recorded the St. Marys wetlands at water levels that varied over a range of 1.43 meters (176.09 to 177.52 m, above the International Great Lakes Datum 1955). This range allowed an analysis of the influence of water levels on the wetlands over a large part of their historic range.

Wetland areas were quantified for each year of aerial photographs by the USFWS National Wetland Inventory contractor (St. Petersburg, FL) experienced with photo interpretation and identification of wetlands. The study area was also visited by the USFWS contractor. Initial interpretations were completed with an analog stereo plotter using the 1984 images. The plotter established

Table 13.1. Dates, Scales, Emulsions and Monthly Average Water Level Elevations of Aerial Photographs

Month	Year	Scale	Emulsion	Monthly Level (m)	Annual Level (m)
July	1939	1:20,000	black & white	177.01	176.62
August	1953	1:16,000	black & white Infrared	177.52	177.19
July	1964	1:16,000	black & white	176.09	176.16
June	1978	1:12,000	black & white	176.74	176.84
October	1982	1:58,500	color infrared	176.82	176.67
September	1984	1:12,000	color infrared	177.18	177.00
October	1985	1:24,000	natural color	177.43	177.20

geometric control and was used to correct any spatial inaccuracies found in the aerial photos. The interpreter viewed stereo pairs of photos, and outlined the wetland boundaries. The other dates of photography were compared to the 1984 maps and boundary adjustments made with a Zoom Transfer Scope. The boundaries were plotted on USGS 7½ minute maps of the study area.

Wetlands up to 800 meters inland from the river shore were included in the analysis to incorporate wetlands that could be influenced by changing water levels. This determination was based on USGS map elevations and mapped marsh areas. The photo interpreter assigned each wetland area to a type or class following the USFWS Classification System (Cowardin et al., 1979; U.S. Fish and Wildlife Service, 1987). Data were summarized in wetland maps and tables of wetland areas reported by the USFWS Wetland Analytical Mapping System (WAMS). The USFWS used its GIS system to create GIS files of the wetland polygons in ELAS format, which could be read by ERDAS software (U.S. Fish and Wildlife Service, 1988; Williams and Lyon, 1991).

The National Wetland Inventory provided maps and WAMS-derived tabular summaries. These were appropriate for completing some analyses. Further GIS analysis using the digital files, however, offered several advantages. First, the numerous wetland classes identified by National Wetland Inventory could readily be aggregated to simplify the analyses and to employ wetland classes used in comparable wetland studies on the Great Lakes. This was effective because of the large number of wetlands found on five USGS quadrangles and the aggregation was accomplished rapidly using the ERDAS software program RECODE. Other data sources on the St. Marys wetlands indicated that the Unconsolidated Bottom class included larger areas with submergent vegetation cover than the USFWS definition of Unconsolidated Bottom (less that 30%). Much of the bottom of the St. Marys is vegetated (Liston et al., 1986), hence the Unconsolidated Bottom class included Aquatic Bed Wetland (Cowardin et al., 1979).

Because the tabular National Wetland Inventory summaries were completed on individual quadrangles, without a GIS it was necessary to sum changes in five separate quadrangles to determine change across the entire study area. Consequently, the GIS was used to combine the quadrangle areas into a single digital file to reduce the number of analyses. Six digital files (one quadrangle was divided into two digital files) from each year were combined into one mosaic. This provided seven mosaics of the study area corresponding to the seven years of historical aerial photographic coverage: 1939, 1953, 1964, 1978, 1982, 1984, and 1985. The mosaics were created using the ERDAS GIS software program SUBSET.

There were also slight discrepancies in the total study area measured in both the tabular data and the individual digital files. Small areas were included in some years that were not in others. The GIS was used to establish an exact common area for each year so that year to year changes in wetland class and area could be determined more accurately. A GIS modeling package (GISMO) was used to create a mask, a sum of areas that were not common to each of the seven files, that was applied to each mosaic so the study area was precisely the same for each of the seven years. Data on the areal extent of each wetland class for each year were then obtained from the corrected mosaics (ERDAS BSTATS). This corrected discrepancies in total area between the seven data sets of up to 1%. Discrepancies in the ranges of the five wetland classes were reduced by as much as 3.8%. Rather than National Wetland Inventory tabular data, these quantities were then used in further analysis (Table 13.2).

The GIS analysis also allowed for the spatial location of changes in wetlands. Analysis of tabular or other data did not allow such identification. Locating exactly where a change occurred was important to identifying other possible causative factors for area changes such as vessel traffic in adjacent commercial navigation channels.

Locational information can also be used with GIS cross-tabulation algorithms to examine the nature of wetland class changes. A transition matrix for the St. Marys River study is shown as

Table 13.2. Quantities of Wetland Classes Calculated by GIS from Mosaics (Hectares). Note: Mask included Upland

Year	Unconsolidated Bottom	Emergent Wetland	Unconsolidated Shore	Scrub Shrub	Forested Wetland
1939	5097.4	1248.7	348.7	352.5	178.7
1953	5586.7	924.7	3.1	321.1	272.3
1964	5100.5	1326.5	99.2	384.3	250.3
1978	5611.2	899.8	9.2	341.4	281.2
1982	5554.5	950.5	7.9	333.1	283.2
1984	5601.1	909.3	1.6	339.9	281.7
1985	5600.8	913.5	1.5	341.7	282.2

Table 13.4. The matrix shows the net area changes that took place between wetland classes from one wetland state to the next. For example, the upper left cell in Table 13.4 indicates that 174.6 ha of Unconsolidated Bottom were gained in 1953 that had been Emergent Wetland in 1939. This matrix was used to examine some of the wetland class dynamics caused by fluctuations in the levels of the St. Marys River.

Class by Class Analysis of Area Changes Due to Long-Term Water Level Changes

The class areas calculated from the six-quadrangle mosaics of the study areas were used for analysis of the effects of water level fluctuations on the wetland class areas. Regressions of class areas on water level were run for the wetland classes Unconsolidated Bottom, Emergent Wetland, Unconsolidated Shore, Scrub-Shrub Wetland, and Forested Wetland. Initially, monthly mean water level from the U.S. Slip gauge was used as the independent variable. Because wetlands do not respond immediately to water level change, and area changes due to water level fluctuations may take months or years to be completed, another set of regressions was run using mean annual water level. Variations in the independent variable, average annual water level, for 1925 through 1985 from the U.S. Slip gauge below the locks at Sault Ste. Marie are shown in Figure 13.2.

Accounting for Lag Times in Wetland Area Changes

Many observations on Great Lakes coastal wetlands (e.g., Jaworski et al., 1979; Busch and Lewis, 1984; Painter et al., 1988) have shown that wetland changes due to the rise or fall of water level may take several years to be completed. To gain some insight into how rapidly wetland areas might have adjusted to water levels, a series of regressions was run using moving average water levels instead of monthly or annual average levels. These regressions were weighted average water levels consisting of the annual water levels of the years of the aerial photographs and annual levels from years prior to the photographs. The levels of the year the wetlands were photographed was given the greatest weight and the preceding years less and less weight. For a two year average the weights were 0.2, 0.1; for a three year average 0.3, 0.2, 0.1, etc.

The standard linear regression equation used for annual mean water levels was:

$$Y = a + b\,(X_{ave}) \tag{1}$$

where Y = wetland area and X_{ave} = average annual water level.

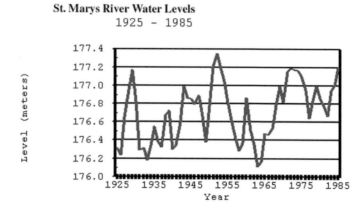

St. Marys River Water Levels
1925 - 1985

Figure 13.2. Mean annual water levels from the U.S. Slip gauge near the study area (meters above ILGD '55).

For weighted average water levels the regression equation was

$$A = a + b\,(c_1{}^*X_t + c_2{}^*X_t - 1 + \ldots + c_n{}^*X_{t-n}) \tag{2}$$

where c_1, c_2...c_n were weighting factors for the average annual water levels of the year of photography, the year before the photography, etc., and the weighting factors were 0.2 and 0.1 for a two year regression, 0.3, 0.2, and 0.1 for a three year regression, etc.

The coefficient of determination (R^2) of the regression equation was maximized by adding yearly increments into the linear combination of levels. An optimal combination of water levels was selected for each wetland class based on the maximized R^2. This optimum fit provided an estimate of the number of years that might have been relevant to the adjustment of each wetland class in the St. Marys River to the water level changes, and therefore the importance of the water levels in the years preceding the measured growing season.

RESULTS

Emergent Wetlands

For Emergent Wetland within the study area there was a 426 ha difference between the year with the largest area, 1964, and the year with the least area, 1978. This represented a 32% change, hence Emergent Wetland area varied by about one-third (Table 13.3). The linear regression of Emergent Wetland area on monthly average water level was not significant. The regression of area on average annual level was significant (* $P < 0.05$). This provided strong evidence of a relationship between water level and area of Emergent Wetland areas between 1939 and 1985. As water levels rose, Emergent Wetland area decreased, and as water levels fell, Emergent Wetland area increased. A number of other studies on Great Lakes emergent wetlands have shown this result (Jaworski et al., 1979; Harris et al., 1981; Enslin and McIntosh, 1982; Lyon and Drobney, 1984; Busch and Lewis, 1984; Greene, 1987; Bukata et al., 1988; Payne et al., 1985; Quinlan, 1985). The fact that the regression of area on monthly average level was not significant and the regression of area on annual average level was significant (** $P < 0.05$, $R^2 = 0.64$) suggested that the effect of water level on Emergent Wetland area took a growing season or longer to complete.

Table 13.3. Direction, Rate, and Magnitude of Wetland Class Changes

Wetland Class	Regression Results			Area Change	
	±	P <	Yr. Lag	Hectares	Percent
Emergent	–	0.001	14	426	32
Unconsolidated Bottom	+	0.001	15	514	9
Unconsolidated Shore	–	0.05	22	347	99
Scrub-Shrub	–	0.01	5	63	16
Forested	+	0.01	22	104	37

There is a tendency for the water levels to be higher in the years of photography taken later in the growing season (Table 13.1). This cannot account for the differences in wetland area, however, because the dominant St. Marys emergent plants reach their maximum heights by July, and plants of maximum height are still present in September and generally October (Liston et al., 1986). *Scirpus acutus* Bigelow shoots persist through the winter and are clearly visible protruding through the ice.

Regressions were run for weighted annual water level averages incorporating water levels from 2 to 30 years previous to the years of the area measurement. The regression fit (R^2) increased for each additional year accumulated through the fourteenth year. The analysis thus suggested that water levels in any given year influenced Emergent Wetland area for up to 14 years, or conversely, that Emergent Wetland area was influenced by water levels of multiple years before its occurrence. At 14 years, the regression accounted for 99% of the variability ($R^2 = 0.991$), and was statistically significant (*** $P < 0.001$).

The question of response time for changes in Emergent Wetland has had less attention. Several investigators have postulated that it takes around five years for Emergent Wetlands to adjust to water level changes. Busch and Lewis (1984) used a five-year weighted moving average to explain changes in a marsh in eastern Lake Ontario marsh, and Painter et al. (1988) used a similar method to explain changes in Cootes Paradise on the western end of the same lake. The increase in R^2s for regression using moving averages up to 14 years suggest that the Emergent Wetlands in the St. Marys River might take even longer to adjust to water levels. There is some rationale for supporting the hypothesis that these wetlands may on average take five years or more to complete adjustments to major water level fluctuations. It was clear from the seven sets of aerial photographs that distribution patterns of *Scirpus acutus* Bigelow were very stable. Some areas appeared not to have changed significantly from 1939 to 1985. Although they are not woody plants, many of the emergents in the St. Marys (e.g., *Eleocharis smallii* Britton, *Sparganium eurycarpum* Engelm., *S. chlorocarpum* Rydb., and *Scirpus americanus* Pers.) are long-lived perennials. Storage effects from low levels can occur as gains in colonized areas and increases in robustness and storage products in rhizomes. Substrate (Lyon et al. 1986) and topographic changes brought about by the water level changes may also take some time to complete.

The regressions took into account both gains and losses, and it is common sense to assume that losses of Emergent Wetland could occur quite rapidly in response to high water levels, certainly less than 14 years. Nonetheless, observations of cattail (*Typha* sp.) responses to water level increases at Horicon Marsh in the state of Wisconsin, USA by Mathiak (1971) indicated that maximum die-off occurred in the fourth and fifth years after exposure to high water levels. Given that some perennial marsh species are resistant to flooding, and that the resistance is likely to vary from species to species, it is probable that responses to increased levels will vary from marsh to marsh depending upon species composition. There appears to be some evidence for this variation (Williams, 1995).

The transition matrix shows some of the interclass transfers that took place between Emergent Wetland and other classes as levels changed. There were large changes from Unconsolidated Bottom to Emergent Wetland associated with the drops in water level from 1953 to 1964 and 1978 to 1982. There were also large transfers from Emergent Wetland to Unconsolidated Bottom during the level increase from 1939 to 1953 and 1982 to 1984. There were fairly large losses of Emergent Wetland in the 1939 to 1953 transitions to Scrub-Shrub, Forested Wetland, and Upland. These would not be expected from water level changes and were probably due to succession that occurred in spite of the level increases.

Unconsolidated Bottom

There was a 514 hectare change in area of Unconsolidated Bottom between the year with the largest area, 1978, and the year with the smallest area, 1964. This amounted to a 9% change in area (Table 13.3). Unconsolidated Bottom area responded in a manner opposite to that of Emergent Wetland. Unconsolidated Bottom expanded during high water levels and contracted during low water levels; as water levels increased, Unconsolidated Bottom area increased. The regression of Unconsolidated Bottom area on monthly water levels was not statistically significant. The regression of area on annual average level was significant (** $P < 0.05$, $R^2 = 0.55$) and suggested that the effect of water level on Unconsolidated Bottom area also took a growing season or more to complete.

A series of regressions was run for weighted average water levels for 2 to 30 years previous to the years of area measurement. The regression R^2 increased for each additional year up to 15 years. The analysis thus suggested that it takes several years before the effects of water levels on Unconsolidated Bottom are complete. At 15 years, the regression (Table 13.3) accounted for 99% of the variability ($R^2 = 0.995$), and was statistically significant (*** $P < 0.001$). The similarity in lag time (15 vs. 14 years) and the opposite response suggested a reciprocal connection between Unconsolidated Bottom and Emergent Wetlands. As water levels increased, many of the losses in Emergent Wetland area were probably attributable to gains in Unconsolidated Bottom and as water levels decreased, the opposite exchanges probably occurred. This was supported by the information in the transition matrix in Table 13.4 discussed above.

Unconsolidated Shore

Unconsolidated Shore is by definition wetland with less than 30% coverage by vegetation and a hydrologic regime varying from intermittently flooded to intermittently exposed. The area of Unconsolidated Shore changed by 347 ha between its minimum and maximum. It varied by 99% of its maximum value (Table 13.3). There was a general decreasing trend until it nearly disappeared in 1984 and 1985. The data suggest that there was gradual colonization of areas that were nearly bare in 1939, possibly because of some disturbance. Regressions of area of Unconsolidated Shore on both monthly and mean annual water level were not significant.

As the regressions on weighted annual mean levels included more and more years prior to the dates of the photographs, the regressions became significant. The results suggested that water levels as early as 22 years prior to the date of the wetland area measurements may have had an influence on the amount of Unconsolidated Shore wetland. The R^2 of the regression reached a maximum of 0.75 with a weighted average that included 22 years of average annual levels prior to

Table 13.4. Transition Matrix of Wetland Classes Transfers Developed from GIS Cross Tabulation[a]

	1939>1953	1953>1964	1964>1978	1978>1982	1982>1984	1984>1985
			Changes in Unconsolidated Bottom			
EM	174.6	−355.6	393.1	−60.5	39.6	−1.8
US	287.8	−77.7	85.5	2.3	4.9	−0.3
SS	18.5	−28.3	24.9	3.3	1.8	−0.0
FO	0.4	2.1	−0.9	1.0	0.1	−0.1
UP	7.7	−26.5	7.9	−2.8	0.2	2.0
			Changes in Emergent Wetland			
UB	−174.6	355.6	−393.1	60.5	−39.6	1.8
US	24.5	−10.9	2.3	0.1	0.5	−0.0
SS	−64.4	−2.4	−13.9	−6.7	−3.5	0.1
FO	−48.9	29.1	−10.5	2.1	0.5	0.3
UP	−60.4	30.2	−11.2	−5.4	0.9	2.0
			Changes in Unconsolidated Shore			
UB	−287.8	77.7	−85.5	−2.3	−4.9	0.3
EM	−24.5	10.9	−2.3	−0.1	−0.5	0.0
SS	−11.7	1.5	1.1	0.1	0.5	−0.8
FO	−1.2	0.1	−0.0	−0.0	−0.1	0.0
UP	−20.3	5.9	−3.2	0.9	−1.2	0.4
			Changes in Scrub-Shrub Wetland			
UB	−18.5	28.3	−24.9	−3.3	−1.8	0.0
EM	64.4	2.4	13.9	6.7	3.5	−0.1
US	11.7	−1.5	−1.1	−0.1	−0.5	0.8
FO	−59.3	3.8	−11.2	−4.7	3.8	0.1
UP	−29.8	30.2	−19.6	−6.9	1.7	1.0
			Changes in Forested Wetland			
UB	−0.4	−2.1	0.9	−1.0	−0.1	0.1
EM	48.9	−29.1	10.5	−2.1	−0.5	−0.3
US	1.2	−0.1	0.0	0.0	0.1	0.0
SS	59.3	−3.8	11.2	4.7	−3.8	−0.1
UP	−15.3	13.1	8.3	0.3	3.0	0.7

[a] Changes in Hectares

the area measurements (* P < 0.05). Like Emergent Wetland and Scrub-Shrub wetland, Unconsolidated Shore decreased with increasing water levels and increased with decreasing water levels.

The transition matrix (Table 13.4) shows that 288 ha of the dramatic loss in area in 1939 went to Unconsolidated Bottom in 1953. This occurred simply because the water levels went up; no vegetation change was necessary. There was also a large loss to Unconsolidated Bottom with the level increase from 1964 to 1978. Table 13.4 also shows significant losses of Unconsolidated Shore to Emergent, Scrub-Shrub, and Upland in the 1939 to 1953 transition. These suggested changes due to succession rather than water level.

Because the largest interchanges were between Unconsolidated Shore and Unconsolidated Bottom, which occurred instantaneously with water level changes, the long lag period shown in the regression analysis does not appear reasonable. The regression correctly shows a reciprocal relationship between area and water level, but succession evidently played a significant role and complicated the water level—area relationship.

Shrub-Scrub Wetland

Scrub-Shrub Wetland area varied by 63 ha, about 16% of its maximum area (Table 13.3). The changes in Scrub-Shrub Wetland were clearly influenced by water level as the regressions of both monthly and annual average water levels on area of Scrub-Shrub were significant ($*$ $P < 0.05$). Scrub-Shrub wetland area, like Emergent and Unconsolidated Shore areas, decreased during higher water levels and increased during lower water levels.

Changes in Scrub-Shrub Wetland area appeared to develop rapidly in response to water level changes. The R^2 reached a maximum with the regression that included only five years of weighted average water levels ($R^2 = 0.86$). The relationship was statistically significant ($**$ $P < 0.01$; Table 13.3). This suggested that it took several years for the St. Marys River Scrub-Shrub Wetland vegetation to complete its expansions and contractions in response to water level changes, possibly not as long as Unconsolidated Bottom or Emergent Wetland.

The transition matrix shows that there were losses to Unconsolidated Bottom during the 1939 to 1953, and 1964 to 1978 water level increases. There were also changes attributable to succession: the 59.3 ha loss to Forested Wetland, the 29.8 ha loss to Upland, and the 64.4 ha gain from Emergent Wetland. The same pattern of losses to Forested Wetland and Upland and gains from Emergent Wetland is found in the 1964 to 1978 transition. Successional processes apparently have a significant effect in the Scrub-Shrub changes.

Forested Wetland

Forested Wetland varied by 104 ha from maximum to minimum area, a change of 37%. The regressions of area on monthly and annual average water levels were not significant.

The R^2s of the regressions using Forested Wetland area on weighted average water levels increased (to 0.64) as yearly average levels were added through 22 years prior to the date of the area measurements. With this lag, the regression was significant ($**$ $P < 0.01$). This suggested that water levels in any given year appeared to have had an effect on area of Forested Wetlands for as many as 22 years after they occurred. The regression was positive, i.e., Forested Wetlands appeared to increase during periods of high water levels.

The latter outcome was not expected because woody vegetation is generally thought to be less tolerant to flooding than herbaceous vegetation (Keddy and Reznicek, 1985). Wetland trees are more tolerant than Upland trees, however. An increase in water levels can indeed result in an increase in the quantity of Forested Wetlands. While standing water or saturated soil will often kill upland trees, many of the facultative wetland plants will, according to their classification, grow in the presence or absence of standing water or saturated soils. Higher water over time will influence the presence of facultative wetland trees, and allow their increased growth as compared to facultative upland trees. The species found near the St. Marys River are typically facultative wetland species. Moreover there is evidence for this from the 8.3 ha change from Upland to Forested Wetland in the 1964 to 1978 transition. It is notable that Forested Wetland areas remained consistent despite higher water levels in 1984 and 1985.

The gains during low to high transitions can also in part be explained by succession. Table 13.4 indicates that the largest gain in the 1939 to 1953 transition is from Scrub-Shrub (59.3 ha), followed by Emergent Wetland (48.9). The same pattern of gains is found in the 1964 to 1978 "high to low" transition; gains are mostly from Scrub-Shrub (11.2 ha) and Emergent Wetland (10.5 ha).

The Forested Wetland changes from 1953 to 1964 were puzzling, however. Indeed, 13.1 ha of 1953 Forested Wetland area came from Upland, and 29.1 ha were lost to Emergent Wetland, the opposite of succession. An explanation may be found in the timing of the large water level increase in that it was affecting the 1953 wetlands. Figure 13.2 shows that St. Marys River levels were in the third year of a high level spike that lasted four years. That spike could have caused some areas of Upland to change to Forested Wetland and also caused the ultimate loss of some Forested Wetland area to Emergent Wetland and Scrub-Shrub, but not so rapidly as to be visible in the 1953 photographs. The loss may have been either not evident , i.e., the trees were dead but still standing, or the loss may have occurred in the next few years. The successional changes did not show up until after 1953, however.

The regression data for Forested Wetland is at best ambivalent. The transition matrix indicates that Forested Wetland increases occur during transitions from low to high levels as successional processes, i.e., conversion of Emergent Wetland and Scrub-Shrub to Forested Wetland. These are not driven by high water levels, hence the association of larger Forested Wetland areas with higher levels shown in the regression is probably an artifact of the timing of successional processes. Large areas of Emergent Wetland and Scrub-Shrub Wetland were evidently caused by the long period of low water in the 1930s (Figure 13.2). These in part "succeeded" to Forested Wetland despite the high levels in the early 1950s.

DISCUSSION

Because a GIS is a convenient tool for manipulation of mapped data, it simplified development of data sets that could be used in data analysis. Of particular interest in this context was the use of a transition matrix derived from a GIS cross tabulation. The matrix showed interclass transfers from wetland state to wetland state. In this study, the transitions matrix was particularly useful in helping to identify successional effects and distinguish them from water level effects.

If the R-Squares of the regressions were taken as a measure of how much variability in the data is due to the regression, the R-Squares then provided a measure of how much of the variability in each wetland class area was due to changes in water levels. This suggested that virtually most of the variability ($R^2 = 0.99$) in the areas of both Unconsolidated Bottom and Emergent Wetland was due to water level fluctuations. A significant part of the variability in Scrub-Shrub Wetland was also due to level fluctuations in the St. Marys ($R^2 = 0.86$). By statistical measures, there were apparently other important factors affecting Unconsolidated Shore ($R^2 = 0.74$) and Forested Wetland ($R^2 = 0.78$).

The transition matrix was useful in identifying strong evidence of successional trends in these wetland classes by allowing quantification of interclass transfers. The matrix also provided a basis for suggesting that the association found in regression between high water levels and increased Forested Wetland was probably anomalous.

There are several factors that could not be accounted for in this study that may have affected conclusions concerning the wetland class area changes. Accuracy of photo interpretation by the USFWS contractor may have affected data accuracy. There may also have been human influences on area changes. Some of these shores may at one time have been farmed; Forested Wetlands may have regrown in areas previously cut down for agriculture. Filling in some areas was visible from changes in the photographs between 1953 and 1978, causing losses of wetland not due to level fluctuations.

CONCLUSION

Use of GIS and Historical Aerial Photography

Using a GIS for a wetland change detection study proved to have several advantages over manual methods for use of historical photographs, such as mapping and measuring areas with a plane planimeter and use of overlays. Some of the data in this study could have been derived from the tabular data supplied by the National Wetland Inventory. The tabular data was in effect a National Wetland Inventory GIS product however, and would have been produced from the authors' GIS, had the mapping and digitization not been completed by the National Wetland Inventory.

GIS use allowed creation of mosaics of all study USGS quadrangles, greatly reducing the number of necessary analytic operations. It provided a means of quickly establishing an accurate common area of study for each year of wetland mapping. It provided a means of rapidly aggregating wetland types to convenient wetland classes. It provided an efficient method of measuring historical wetland class changes based on archived aerial photographs. The change measurements could then be used in conjunction with the historical water level data to understand how water level changes affected wetland class changes between 1939 and 1985.

Finally, GIS use allowed rapid development of locational information. Digital maps of areas in which Emergent Wetlands were lost provided information on potential losses besides water levels, such as the proximity of commercial navigation channels. Cross tabulation provided the capability to develop a transition matrix that allowed quantification of class to class changes between wetland states in 1939, 1953, 1964, 1978, 1984, and 1985. These data were particularly useful in confirming expected change patterns and in explaining changes that were not intuitive. Hence this study indicates that GIS and historical aerial photography can be used effectively in tandem to study how water level changes affect wetland class areas. These properties of Great Lakes wetlands are important to know, not only because they provide a better understanding of wetland processes, but also in view of the prospect of Great Lakes water level regulation changes, or the adoption of more extensive regulation of the levels of the Great Lakes and connecting channels.

REFERENCES

Bukata, R.P., J.E. Bruton, J.H. Jerome, and W.S. Haras, 1988. A Mathematical Description of the Effects of Prolonged Water Level Fluctuations on the Areal Extent of Marshlands. Scientific Series No. 166, Inland Waters Directorate, National Water Research Institute, Canada Center for Inland Waters, Burlington, Ontario.

Busch, W.D., and C.N. Lewis, 1984. Responses of Wetland Vegetation to Water Level Variations in Lake Ontario. *Proceedings of the Third Annual Conference on Lake and Reservoir Management,* North American Lake Management Society, pp. 519–523. U.S. Environmental Protection Agency, Washington, D.C.

Cowardin, L.M., V. Carter, F.C. Golet, and E.T. LaRoe, 1979. Classification of Wetlands and Deepwater Habitats of the United States. U.S. Department of the Interior, Fish and Wildlife Service, Report No. FWS/OBS–79/31. Washington, D.C.

Enslin, B., and D. McIntosh, 1982. Changes in Aquatic Vegetation in Qanicassee, Nayanquing Point and Wildfowl Bay. Center for Remote Sensing, Michigan State University. Prepared for the East Central Michigan Planning and Development Region.

Greene, R.G. 1987. Effects of Lake Erie Water Levels on Wetlands Areas Measured from Aerial Photographs: Pointe Mouillee, Michigan. M.S. Thesis. Department of Civil Engineering, Ohio State University, Columbus, OH.

Harris, H., G. Fewless, M. Milligan, and W. Johnson, 1981. Recovery processes and habitat quality in a freshwater coastal marsh following a natural disturbance. In *Selected Proc. Midwest Conf. on Wetland Values and Man*, pp. 363–379. Freshwater Society, St. Paul MN.

Jaworski, E., C.N. Raphael, P.J. Mansfield, and B.B. Williamson, 1979. Impact of Great Lakes Water Level Fluctuations on Coastal Wetlands. Office of Research Development, Institute for Water Research, Michigan State University, East Lansing, MI.

Keddy, P.A., and A.A. Reznicek, 1985. Great Lakes vegetation dynamics: The role of fluctuations water level and buried seeds. *J. of Great Lakes Res.* 12:25–36.

Liston, C.R., and C.D. McNabb, (principal investigators) with D. Brazo, J. Bohr, J. Craig, W. Duffy, G. Fleischer, G. Knoecklein, F. Koehler, R. Ligman, R. O'Neal, M. Siami, and P. Roettger, 1986. Limnological and Fisheries Studies of the St. Marys River, Michigan, in Relation to Proposed Extension of the Navigation Season, 1982 and 1983. Submitted to the Detroit District, Corps of Engineers, Detroit, Michigan and U.S. Fish and Wildlife Service, Twin Cities, MN. Issued as U.S. Fish and Wildlife Service, Office of Biological Services Report OBS/85(2).

Lyon, J.G., and R.D. Drobney, 1984. Lake level effect as measured from aerial photos. *ASCE Journal of Surveying Engineering*, 110: 103–111.

Lyon, J.G., R.D. Drobney, and C.E. Olson, Jr., 1986. Effects of Lake Michigan water levels on wetland soil chemistry and distribution of plants in the Straits of Mackinac. *J. Great Lakes Res.*, 12: 175–183.

Mathiak, H.A., 1971. Observations on Changes in the Status of Cattails at Horicon Marsh, Wisconsin. Research Report 66, Department of Natural Resources, Madison, WI.

Painter, D.S., K.J. McCabe, and W.L. Simser, 1988. Past and present limnological conditions in Cootes Paradise affecting aquatic vegetation. National Water Research Institute, Environment Canada, Burlington, Ontario. NWRI Contribution No. 88–47, 30 pp. plus figures.

Payne, F.C., J.L. Schuette, J.E. Schaeffer, J.B. Lisiecki, D.P. Regalbuto, and P.S. Rogers, 1985. Evaluation of Marsh Losses: Maisou Island Complex. Prepared for Wildlife Division, Michigan Department of Natural Resources.

Quinlan, C.M., 1985. The Effects of Lake Level Fluctuations on Three Lake Ontario Shoreline Marshes. M.A. Thesis, University of Waterloo, Ontario, Canada.

U.S. Army Corps of Engineers, Detroit District, 1985. Great Lakes Water Level Facts.

U.S. Fish and Wildlife Service, 1987. Photointerpretation Conventions for the National Wetlands Inventory. May 1, 1987.

U.S. Fish and Wildlife Service, 1988. St. Marys River, Michigan, Geographic Information Systems Applications. Final Report, January 19, 1988.

Williams, D.C., 1995. "Dynamics of Area Changes in Great Lakes Coastal Wetlands Influenced by Long Term Fluctuations in Water Levels. Ph.D. Dissertation, School of Natural Resources and Environment, University of Michigan, Ann Arbor, MI.

Williams, D.C., and J.G. Lyon, 1991. Use of a geographic information system data base to measure and evaluate wetland changes in the St. Marys River, Michigan. *Hydrobiologia*, 219: 83–95.

CHAPTER 14

Watershed-Based Evaluation of Salmon Habitat

Ross S. Lunetta, Brian L. Cosentino, David R. Montgomery,
Eric M. Beamer, and Timothy J. Beechie

INTRODUCTION

Dramatic declines in Pacific Northwest (PNW) salmon stocks have been associated with land-use induced freshwater habitat losses (Nehlsen et al., 1991). Substantial resources are being directed toward the restoration of stream habitats in efforts to maintain and/or restore wild salmon stocks in the PNW, yet there is no common scientific framework for guiding the prioritization of where and how salmon habitat preservation and restoration activities should occur. GIS-based analysis can provide a systematic tool for targeting restoration opportunities by rapidly characterizing potential salmon habitat over large geographic regions and by providing baseline data for development of habitat restoration strategies. When integrated, data on stream channels, riparian habitat, and watershed characteristics provide a powerful tool for the development of watershed restoration and management strategies (Delong and Brusven, 1991).

Previous efforts to prioritize salmon habitat preservation and restoration opportunities on state and federal lands in Oregon (Bradbury et al., 1995) and Washington (Oman and Palensky, 1995) have met with some success, but problems associated with data availability over large geographic regions have limited applications. The objectives of this study were to: (a) develop a rapid, cost-effective, and objective analytical tool to support prioritization of specific subbasins and watersheds for salmon habitat preservation and restoration opportunities; (b) investigate the correspondence between forest seral stage and large woody debris (LWD) recruitment and associated pool-riffle stream bed morphologies; (c) illustrate the creation of integrated baseline data to support watershed analyses and the development of preservation and restoration strategies; and (d) explore the use of such data to facilitate the communication of scientific information to decision-makers and the public.

APPROACH

Classification schemes impose order on a system for some particular purpose. Stream channel classifications, for example, provide a means to evaluate and assess the current condition and potential response of channel systems to disturbance (natural and anthropogenic). Identification of functionally distinct channel types can also target fieldwork on stream reaches of particular interest and provide a reference frame for communication between multidisciplinary groups evaluating habitat conditions. No channel classification is ideal for all purposes, and the approach adopted should reflect the goals to which a classification will be applied. Our project needed to identify the likely location and quality of salmon habitat from existing regional data. Numerous channel classification systems rely on the integration of physical variables such as channel slope, channel morphology, and channel pattern (Paustian et al., 1992; Montgomery and Buffington, 1993; Rosgen, 1994).

Several existing channel classifications can be applied to PNW streams. Paustian et al. (1992), for example, broadly classify stream channels according to fluvial process groups (i.e., estuarine, palustrine, alluvial fan, etc.). Rosgen (1994) combined channel slope, cross-section morphology, and plan view morphologic attributes to classify stream reaches into general categories. Rosgen (1994) then included channel entrenchment, width to depth ratio, sinuosity, slope, and bed material to further refine stream type categories. Of these attributes only channel slope is readily determined from typical digital data available over broad regions. Moreover, neither approach allows modification of channel type due to the influence of large woody debris contributed from streamside forests, which can be a primary influence on the morphology of stream channels in the Pacific Northwest (Swanson and Lienkaemper, 1978; Keller and Swanson, 1979; Montgomery et al., 1995; Abbe and Montgomery, 1996).

For this study we selected the classification system of Montgomery and Buffington (1993) which broadly stratifies channel morphology and allows for adjustment of channel type due to morphologic influences of LWD, and can be applied over large areas on the basis of correlations with reach average slope. At the reach level of classification, channel morphology is controlled by hydraulic discharge, sediment supply, and external influences such as LWD. The classification identifies distinct alluvial bed morphologies (e.g., pool-riffle, plane-bed, step-pool), and the influence of large woody debris on "forcing" stream morphology is designated by modifiers added to a particular reach label (e.g., forced pool-riffle). These specific channel morphologies can be generalized into source, transport, and response reaches (Montgomery and Buffington, 1993). In mountain drainage basins, source reaches tend to be debris-flow-prone colluvial channels that function as headwater sources of sediment to downstream reaches. Transport reaches tend to be step-pool and cascade morphology reaches that rapidly convey increased sediment loads to lower-gradient downstream channels. Response reaches are pool-riffle and plane-bed channels that can exhibit dramatic morphologic response to increased sediment loads. Channel reach slope (S) generally correlates with reach morphology, particularly at the coarse level distinctions of source ($S \geq 0.20$), transport ($0.04 \leq S < 0.20$), and response reaches ($S < 0.04$).

Among response reaches, several morphologic types may occur. Channels with slopes between 0.001 and 0.01 typically exhibit pool-riffle morphology regardless of LWD loading levels, whereas channels with slopes between 0.01 and 0.02 are LWD dependent: at low LWD loading, these reaches typically have either a pool-riffle or plane-bed (i.e., riffle-dominated) morphology, whereas at higher LWD loading, LWD pieces and LWD jams force the formation of pools, hence the name forced pool-riffle channel. Channels in the 0.02 to 0.04 slope range typically exhibit plane-bed or forced pool-riffle morphologies, depending upon LWD loading (Montgomery and Buffington, 1997). Channels with slopes above 0.04 typically exhibit step-pool or cascade morphologies. Of these channel types, salmonid species appear to strongly prefer pool-riffle and forced pool-riffle channels.

Identification of response reaches provides a simple method for identification of potential salmon habitat, as the zone of anadromous fish use typically is restricted to these low-gradient reaches in Pacific Northwest watersheds (Montgomery, 1994). Also, the age class of streamside forests can indicate the potential for a source of abundant large woody debris to stream channels. Channel slope can be readily determined from digital elevation models. We coupled this coarse slope-based classification of channel types with remote sensing data on the associated seral-stage of the streamside forest to generate a regional GIS-driven classification of potential salmonid habitat locations and quality. This approach, however, simply provides an indication of the likely channel type and a general sense of the likely woody debris loading. Channel slopes determined from digital elevation models (DEMs) and wood loadings derived from forest seral stage correlations can be misleading due to: (a) poor topographic representation in the DEM at the scale of channels; (b) the natural overlap and range in slope for different channel types; (c) variations in channel type due to local controls; and (d) differences between in-channel LWD loading and riparian forest conditions due to removal of LWD from channels flowing through mid- to late-seral stage forests or clear-cutting of streamside forests without removal of in-channel LWD. In spite of these caveats, the simple classification based on general channel type and riparian forest seral stage provides a direct screening tool for identifying likely sites of low- and high-quality salmon habitat.

We infer that pool abundance correlates with overall habitat quality, and that LWD loading is an important factor in determining habitat quality in response reaches. Assuming that riparian forest conditions correlate with increased LWD loading, it follows that the condition of the adjacent riparian forest would correlate with channel type over the slope range of approximately 0.01 to 0.04, with older forests having higher potential for LWD loading and a higher probability of being forced pool-riffle reaches (high quality habitat). Conversely, reaches with young forests or no forest along the channel have a higher probability of being a plane-bed reach (poor quality habitat).

Application of our results to prioritize specific subbasins and watersheds for salmon habitat protection and restoration efforts is based on three major assumptions: (1) salmon stocks are adapted to local environmental conditions; (2) preservation of "natural" conditions will benefit multiple salmonid species; and (3) a general categorization of channels adequately describes key habitat elements for multiple salmonid species. The first two assumptions are more completely explained by Peterson et al. (1992) and Beechie et al. (1996). The third assumption is supported by limited data showing that several species and life history stages select the same two channel types over others (E. Beamer, unpublished data). These preferences appear to be related to factors such as pool area and depth, cover complexity, and the quality of spawning gravels.

METHODS

Watershed screening was performed at both the subbasin (>450 km^2) and watershed (<260 km^2) scales to identify probable high quality and degraded locations in western Washington State. For the purposes of this study, subbasins correspond to Washington Department of Natural Resources (WDNR) Water Resource Inventory Areas (WRIAs) and watersheds correspond to WDNR Watershed Administrative Units (WAUs). The multiple analytical scales provide comparative evaluations of potential salmon habitat across large geographic regions (e.g., western Washington State) or for evaluations across watersheds within an individual subbasin. Data outputs were summarized by WAU, which typically comprise 120 to 260 km^2. WAUs are subunits within larger subbasins (WRIA) that range from 450 to 6,500 km^2 in western Washington State. The GIS-based predictions of potential habitat locations serve to extend the spatial extent of field observations across the entire study area.

Data Sources

Ideally, all spatial data sources should be derived and used at a scale commensurate with the ecological processes of interest. For this project the appropriate source data scale for watershed analysis and management activities across the western Washington project area is 1:24,000 and larger, to accurately resolve the location of salmonid stream habitat. However, large-scale digital data sets such as vegetation cover and land ownership were not available over the project area. At the expense of spatial resolution, some coarser resolution data sets were used (Table 14.1).

Two sources of digital hydrographic data were available: (1) the U.S. Environmental Protection

Table 14.1. Study Data Sets and Corresponding Scale, Formats, and Source Description

Data	Scale	Format	Description
DEM	1:24,000	raster	7.5-minute; 30-meter cell; Levels 1 & 2.
Hydrography	1:24,000	vector	Compiled from USGS 7.5-minute quads & aerial photography.
Transportation	1:24,000	vector	Compiled from USGS 7.5-minute quads & aerial photography.
WAU Boundaries	1:100,000	vector	Variable accuracy due to multiple regional mapping efforts.
WRIA Boundaries	1:24,000 1:62,500	vector	Boundaries developed by state natural resource agencies in cooperation with the USGS.
Forest Vegetation Seral Stage	~1:100,000	vector/ raster	Landsat Thematic Mapper (TM)-derived forest cover.
Land Use/Land Cover	1:250,000	raster	Non-forested lands (ag./urban/etc.) from USGS Land Use/Land Cover.
Validation Data	1:24,000 1:12,000	Hard copy map/pt. data and tabular data.	Field observations.
Land Ownership	1:100,000	vector	Public land ownership.
Landsat TM	~1:100,000	raster	Terrain-corrected imagery ± 15 meters.

Agency's (EPA's) 1:100,000 scale "river reach files"; and (2) 1:24,000 scale hydrography provided by the WDNR. The river reach files have the advantage of unique identifiers for all stream reach locations. Nonetheless, superior mapping resolution associated with the 1:24,000 scale hydrography data was considered a better match for the needs of this study because most field observations and stream habitat measurements were recorded at scales of 1:24,000 and larger, and absolute stream orientation was critical for subsequent spatial data analyses across multiple thematic data layers.

The only available data source for hydrologic unit delineations at both the subbasin and watershed scales for western Washington State was the WRIA and WAU coverages. WRIA boundaries were compiled from 1:24,000 to 1:62,500 scale maps and WAUs were generally compiled at 1:100,000 scale (WDNR, 1988, 1993). Incongruities between the WAU boundary delineations and the larger scale hydrography were common. For example, along wide river main stems, WAU boundaries were not always in agreement with river main stems, especially as river shape became more sinuous in the 1:24,000 scale hydrography data.

Total road length and road density were important attributes used in the assessment of potential habitat quality at the WAU scale of analysis. Transportation data were available for the project area at 1:100,000 and 1:24,000 scales. The 1:24,000 scale data provided the most complete depiction of primary, secondary, and logging roads.

Available source scales for digital elevation data of western Washington were the 1-degree or three arc-second (~85 meter) data and the 7.5 minute (30 meter) DEM data constructed from 1:24,000 scale maps. Given the need to assess channel slope as accurately as possible for this project, the larger scale data provided the best estimate of stream slope over relatively short stream reaches. The slopes of stream arcs were measured over an arc distance of 150 meters. For this purpose, the 30-meter cell size of the 1:24,000 scale elevation models provided superior topographic resolution.

Table 14.2. Study Land Cover Categories Derived from Landsat 5 Thematic Mapper (TM) Data (PMR, 1993; WDNR, 1994)

Class 1
Late Seral Stage
Coniferous crown cover greater than 70%.
More than 10% crown cover in trees greater than or equal to 21 inches diameter breast height (dbh).

Class 2
Mid-Seral Stage
Coniferous crown cover greater than 70%.
Less than 10% crown cover in trees greater than or equal to 21 inches dbh.

Class 3
Early Seral Stage
Coniferous crown cover greater than or equal to 10% and less than 70%.
Less than 75% of total crown cover in hardwood tree/shrub cover.

Class 4
Other Lands in Forested Areas
Less than 10% coniferous crown cover (can contain hardwood tree/shrub cover; cleared forest land, etc.).

Class 5
Surface Water
Lakes, large rivers, and other water bodies.

Class 15
Nonforest Lands
Urban, agriculture, rangeland, barren, glaciers.

Note:
(1) Forest cover derived from Landsat Thematic Mapper™ satellite imagery.
(2) Class 5 derived from Landsat™ and 1:250,000-scale USGS Land Use/Land Cover data.
(3) Class 15 derived from 1:250,000-scale USGS Land Use/Land Cover data.

Forest cover data were originally derived from 1988 Landsat 5 TM data (PMR, 1993) and updated with 1991 and 1993 TM data using image differencing followed by level slicing to identify new clear-cuts (Collins, 1996). The nominal data resolution of 30 meters was interpolated to 25 meters during the terrain correction process. Standard digital image interpretation techniques were then applied to generate the forest cover data (PMR, 1993). Forest cover was broadly categorized into four classes based on forest type and age class (Table 14.2). The overall thematic accuracy of the 1988 TM-based land cover categorization was 92% (PMR, 1993).

The nonforest land cover and most surface water features were derived from 1:250,000 scale U.S. Geological Survey land cover/use data. The data were overlaid on the forest cover classification to discriminate nonforest lands, such as agriculture and urban areas, from forest lands (PMR, 1993). Thus, the final land cover layer contained a mixture of source scales ranging from approximately 1:100,000 to 1:250,000.

Field data used to validate stream channel type prediction were provided as part of an ongoing salmon habitat inventory and management effort. Inventory efforts focused primarily on streams with relatively low channel slopes (<4.0%) and were compiled on 1:12,000 scale orthophotos. Channel slope data were collected using either transit, hand level, or clinometer measurements.

Data Quality and Error Propagation

Errors associated with remote sensing and GIS data acquisition, processing, analysis, data conversion(s), and final data presentation can significantly impact the confidence associated with resultant products and thus influence their utility in the decision-making process. It was not feasible within the scope of this project to explicitly measure discrete error sources and calculate an error propagation budget. Thus, channel type prediction was the only accuracy assessment performed. Channel type accuracy reflects key input data limitations and processing errors. The final GIS data products may give the appearance of uniform thematic accuracy; however, there may be significant variability across specific geographic locations based on the least accurate input data source (Lunetta et al., 1991).

Spatial Data Preprocessing

Data Format Conversions

Data conversion from vector to raster can cause undesirable shifts of objects in the output raster data as well as changes in area and shape (Congalton and Schallert, 1992). This error source was minimized as much as possible by maintaining data in their native format and thus performing limited data conversions. Raster to vector conversions were not performed as part of the project; however, the forest cover data, originally processed from Landsat TM digital imagery were converted to a vector representation prior to processing (PMR, 1993). Also, all single line hydrographic arcs underwent vector to raster conversion to optimize stream buffer calculations (see Stream Buffer Vegetation Tabulation).

Data Generalization

With the exception of the land cover layer, most data sources were not generalized. The Nonforest (class 15) and Surface Water (class 5) classes listed in Table 14.2 were originally compiled under USGS mapping guidelines. The land use and land cover data were interpreted from aerial photography at a scale of 1:60,000 or larger and compiled on 1:250,000 scale topographic maps (USDI, 1993). The guidelines specify a 4.0 hectare minimum mapping unit for urban/built-up lands, surface water, and some agricultural areas. The minimum mapping unit for cropland, pasture, and barren lands is 16.2 hectares. As noted above, the nonforest and surface water data were overlaid on the seral stage coverage to create a combined land cover layer. This layer was subsequently converted to vector format.

Prior to conversion of the land cover data to vector format, a filtering procedure was performed on the raster data coverage to merge polygons smaller than nine pixels into adjacent polygons using a simple majority rule decision criteria. Subsequently, vector polygons smaller than the minimum mapping unit size of 2.0 or 4.0 hectares (depending on adjacent land cover type) were removed (PMR, 1993). Thus, stream bank forest land cover was not accurately represented for patch sizes of less than 4.0 hectares.

Geometric Rectification

GIS processing of multiple data layers requires that all layers reside in a common map projection. A common projection was determined prior to processing based upon possible error sources and processing efficiency. Because project outputs were assumed to be most sensitive to elevation errors, the DEM data were maintained in their native Universal Transverse Mercator (UTM) projection. Changing their projection would have introduced additional error and increased data processing time due to interpolation of cell values as both the DEM and vector land cover data existed in UTM space. Therefore all input data not in UTM space were projected to the UTM space of the DEMs and land cover data.

DEM Processing

ARC/GRID analytical routines were used to mosaic 7.5-minute DEMs for each WRIA. Once a grid was created for a WRIA, it was next processed to remove "sinks" (sinks are cells with an undefined flow direction). The processed DEM was then used in stream channel slope calculations. Additional DEM processing was performed to create a slope grid for each WRIA for use in summary statistics compilation. The slope grid was then recoded into three landscape slope classes: (a) Class 1, 0–29%; (b) Class 2, 30–65%; and (c) Class 3, greater than 65%.

Preparation of Hydrography

The 1:24,000 scale hydrography data, originally tiled by township, were appended and clipped to the respective WRIA boundary. Stream direction was set to point upstream. Stream percent slope was then computed from DEM values for each arc. Start and end elevation and slope were written to the hydrography coverage arc attribute table.

Forest Seral Stage

This coverage was checked for positional errors and logical consistency by overlaying it with the ancillary geocoded TM data to serve as a base map. The absolute positional accuracy of the TM base map was plus or minus 15 meters (Table 14.1). If positional errors were found, then a simple x,y shift was performed to improve geometric fidelity. Thematic inconsistencies between the vegetation layer and the TM data were not reviewed. Ideally, obvious errors, such as urban encroachment on forest lands, would have been corrected through editing procedures using the TM data as a validation data source. However, resource limitations precluded the inclusion of such editing.

Spatial Data Analysis

Each WRIA was processed individually using identical protocols. The first step was to compile all data inputs for processing, followed by creation of summary data statistics, hard copy maps, and graphics. Summary statistics and data graphics were generated for both the WRIA and WAU hydrographic units. Additionally, validation procedures were performed using data from nine WAUs (Bacon Creek, Illabot, Jackman, Nookachamps, Finney, Hansen Creek, Gilligan, Mt. Baker, and Alder) located within the Upper and Lower Skagit River WRIAs. The categorization of stream channel types was accomplished using an automated procedure to calculate slope for individual stream reaches.

The sampling procedure was initiated at the low elevation end of the arc, and measured upstream the specified sampling distance of 150 meters. If a slope less than 4% was found over the sample distance, then the arc ID number and UTM coordinate at the end of the sample distance were stored in a file. Upon locating a slope less than four percent, the procedure then moved upstream along the arc another sample distance and measures the slope. The process was repeated if a slope less than 4% was found, otherwise the remainder of the current arc was abandoned and the next one is sampled. Stream segments listed in the output file were then split at the specified UTM coordinates using an automated editing procedure. After the editing procedure, the updated slope and elevation values were written to the edited hydrography coverage.

Stream Buffer Vegetation Tabulation

A 30-meter raster stream buffer was generated along both sides of single-arc streams in the hydrographic data layer. The actual cell size implemented to model stream buffers was six meters; thus, the width of each buffer was modeled with five six-meter cells. Each raster buffer was indexed to the vector hydrography line coverage and the percentage of each land cover category (Table 14.2) was written to its respective arc in the arc attribute table. Raster procedures were incorporated to speed up the buffer processing time through the use of rapid cross tabulation procedures between the buffer areas and the raster vegetation layer. For wider streams and rivers which are depicted with double arcs, buffers were extended from each bank, and vector processing procedures were used to summarize vegetation within each stream buffer. Statistics were then generated from the buffer summary tables for each arc.

Summary Reports

WRIA summary reports were organized by WAU and list attributes for streams, vegetation, roads, slope, and land ownership (Table 14.3). Map and bar chart plot files produced from WRIA coverages and summary table attributes were used to plot graphical aids for watershed assessment teams. WRIA maps can be generated on a large format plotter to depict the following themes: (a) response, transport, and source channel types; (b) vegetation classes; (c) transportation networks; (d) slopes; (e) land ownership; and (f) WRIA and WAU boundaries.

Validation of Stream Channel Type Predictions

Validation was performed by comparing field observations with GIS-generated channel type predictions (Lunetta et al., 2001). Field assessment data were provided to the project for the nine Lower and Upper Skagit River WAUs previously listed. The length of the sample reaches generally ranged from 100 to 300 meters, and the midpoint of each reach was delineated on a topographic map. The comparison was accomplished through creation of an error matrix for each WAU (Story and Congalton, 1986). A Kappa coefficient was calculated using discrete multivariate statistical techniques as a measure of the overall agreement between the stream channel type predictions and field observations (indicated as the major diagonal) versus agreement that is contributed by chance (Congalton et al., 1983). The Kappa coefficient was calculated based on the formula given by Hudson and Ramm (1987).

RESULTS

Of the 164,083 km of stream reaches analyzed, 23.2% (38,002 km) were categorized as response reaches (≤4.0% slope), of which, 8.7% (3,302 km) were associated with late seral and 20.7%(7,867 km) with mid-seral stage forest stream vegetation.

Table 14.3. WRIA Summary Report Attributes, Extent, and Description

Attribute	Extent	Description
Streams	WRIA/WAU	Total kilometers and stream density. Stream density and percent by predicted channel reach type.
Seral Stage	WRIA/WAU/Stream buffer	Hectares and percents.
Slope	WRIA/WAU	Hectares and percent of landscape slope in three classes.
Roads	WRIA/WAU	Total kilometers and road density.

Hydrographic Data Scale

Figure 14.1 (a,b, see color section), clearly illustrates the deficiencies of the 1:100,000 scale hydrography stream network (compared to the 1:24,000 scale product) for depicting the actual stream channel network. In the Finney Creek WAU, a total of 490.1 km of stream length are contained in the 1:24,000 scale hydrography compared to 94.8 km in the 1:100,000 scale product. More importantly, the results of the response reach analyses indicate a significant underestimate of response reaches associated with 1:100,000 scale coverage compared to the 1:24,000 scale (43.0 km and 64.9 km, respectively). The smaller scale EPA hydrographic data in addition to lacking resolution in the number of streams, was also deficient in absolute stream orientation detail.

Stream Slope Sampling

Results of the stream slope sampling procedure are presented in Figure 14.2. The optimal sample length corresponds to the maximum stream arc sampling distance that provides the maximum response reach length. Seven sample distances (100, 125, 150, 175, 200, 225, and 300 meters) were evaluated for each of three WRIAs (Figure 14.2) which represented a broad range of physiographic conditions present throughout western Washington State (Lower Skagit, Willapa River, and Lyre-Hoko). The objective was to determine the maximum effective distance to minimize computational requirements, where 100 meters is the minimum feasible sampling length. Sample length must be sufficiently long to capture the inherent variation of the DEM. Short sample distances are ineffective because the elevation change over the sample length is often very low or zero, and exceedingly long sample lengths tend to mask slope changes.

The stream slope sampling procedure enhanced the detection of response reaches located between stream confluences and the base of steep mountain slopes and identified additional response

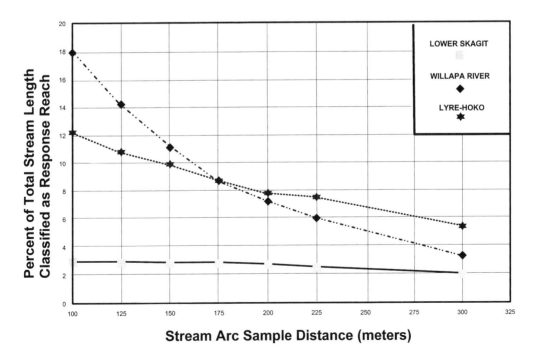

Figure 14.2. Plot of length of predicted response reaches versus sample arc distance as calculated for the Lower Skagit, Willapa River, and Lyre-Hoko WRIAs.

reaches within relatively long stream lengths in moderate terrain. Evaluation of the sample distances indicates that a distance of 100 meters, the shortest distance tested, generated the greatest response length for all WRIAs. As shown for the Willapa WAU (Figure 14.2), the total length of predicted response reaches tends to decrease rapidly as sample length increases from 100 meters to 175 meters, then declines more gradually to 300 meters. The Lyre-Hoko's decline was more gradual than the Willapa, whereas the Lower Skagit was only slightly sensitive to sampling distance. Variations in sampling distance appear to have the greatest effect in locations with moderate to steep terrain. For example, 81% of the landscape of the Lower Skagit WRIA had a slope less than 30 percent; the percentage of area within the Willapa and Lyre-Hoka WRIAs with a landscape slope less than 30% was 75 and 57%, respectively. It appears that hydrologic units with moderate to steep terrain experience the greatest relative increase in response length with decreased sampling distance. Although a 100-meter sample distance maximizes the length of response reaches, a sample distance of 150 meters was applied to minimize errors of commission, and simultaneously reduce processing time and data volume.

Stream Bed Morphology

Field observation data were collected from a total of 120 response reach stream segments in both the Lower and Upper Skagit WRIAs to examine the association between stream buffer zone vegetative land cover and stream bed morphology (Table 14.4). Results indicate that late seral stage forests are associated with forced pool-riffle stream bed morphology. However, the small number of samples (n=8), precludes the drawing of any final conclusions. Response reach buffer zones containing any type of forested land cover had a 77% correspondence to forced pool-riffle stream bed morphology. Nonforested buffer zones were associated with forced pool-riffle morphologies in 35% of the field observations.

Habitat Evaluation

Results applicable to the evaluation of salmon habitat in western Washington State are illustrated in Figure 14.3 (a,b, see color section). The summary bar chart generated for each WRIA and WAU provides a means of comparing potential salmon habitat conditions across WRIAs and to support intra-WRIA assessments. The summary table data for an entire WRIA and individual WAUs include the following information categories: (a) vegetation percent by class; (b) vegetation percent by class within response channel buffers; (c) response, transport, and source channel density; (d) road density; and (e) landscape slope. Summary graphics include drainage density by

Table 14.4. Correspondence between Response Reach Land Cover Categories versus Stream Bed Morphology

Response Reach Land Cover Categories[a]	Percent Reaches Classified as Forced Pool-Riffle
Late Seral Stage	100% (n = 8)
Mid-Seral Stage	78% (n = 18)
Early Seral Stage Other Forest	74% (n = 68)
Non-Forest	35% (n = 26)

[a] observations made along 30m buffers along each bank

channel type and forest seral stage coverage expressed as a percent of total watershed and percent area within buffers around response reaches. These data can facilitate the rapid inference of general streamside conditions and potential for LWD recruitment. In addition, road density and slope data provide some insight to the potential for sedimentation impacts within a given hydrologic unit.

In western Washington more than one-fourth of WRIAs have no late seral stage forest bordering response reaches, and 73% of WRIAs have late seral stage forests along 10% or less of the total response reach length (Figures 14.4 and 14.5). These areas tend to be associated with urban, agricultural or commercial forest land use. Only three WRIAs have more than 20% of their response reach length bordered by late seral stage forests. And these lie partially within national parks or wilderness areas. Overall, only 8.7% of response reaches flow through late seral stage forests. This provides the first quantitative regional characterization of the extent of habitat modifications that accompanied urbanization, agricultural development, and industrial forestry.

Within the Upper Skagit River basin, approximately one-tenth of WAUs had late seral stage forests bordering 10% or less of the total response reach length (Figures 14.6 and 14.7). However, 43% of WAUs had late seral stage forests along 50% or greater of the total response reach length. Of the 20 WAUs identified with late seral stage forests along 25% or greater of total response reach length (highest quality WAUs), eight (40%) were above major dams (Figure 14.7). As in the province scale assessment, land uses in the WAUs with low percent late seral stage tend to be dominated by agricultural and urban development, although some of these WAUs were predominantly industrial forests. WAUs with high percent late seral stage tended to be largely within the boundaries of national parks, national recreation or wilderness areas.

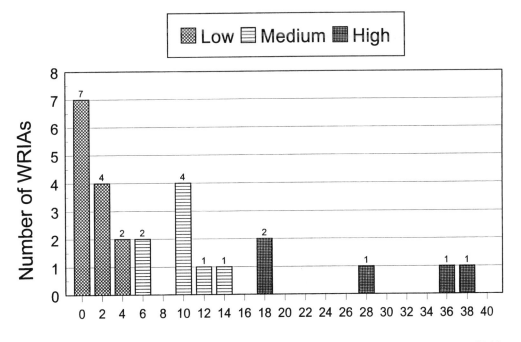

Figure 14.4. Frequency distribution of percent WRIA response reaches in late seral forest stages for western Washington State.

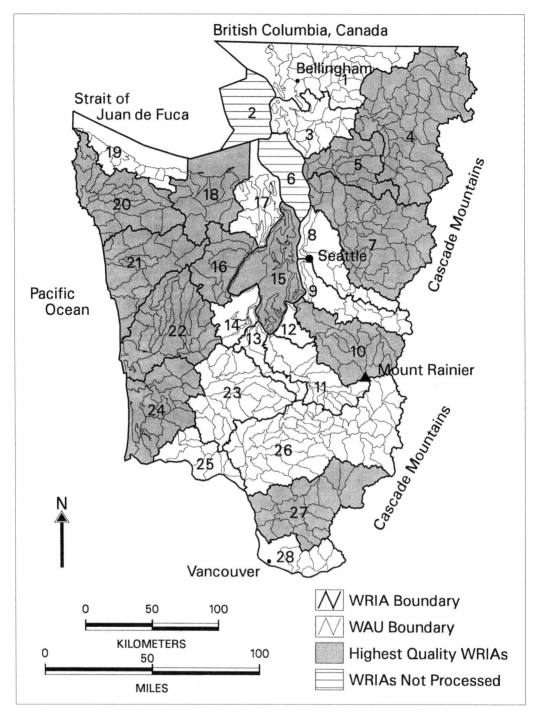

Figure 14.5. Identification and location of the highest quality WRIAs in western Washington State. Note that WRIAs 2 and 6 were not processed because they contain only islands.

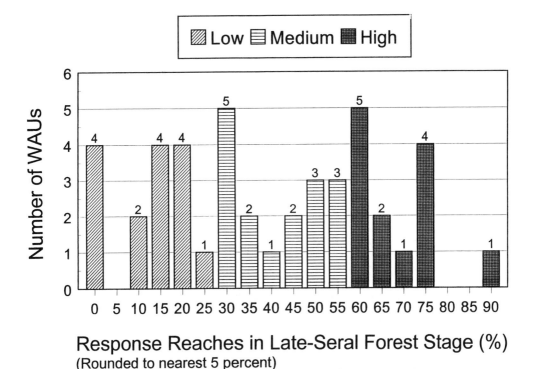

Figure 14.6. Frequency distribution of percent Upper Skagit WAU response reaches in late seral forest stage.

Accuracy Assessment

Although validation was limited to nine WAUs, the basic relationships between physical processes and stream habitat are thought to be consistent across the study area, and the validation for those nine watersheds should be representative for western Washington State. The results of the validation are presented in an error matrix (Table 14.5). The identification of response reaches was 96% accurate, and the overall accuracy of all channel type predictions was 79%, Kappa statistic = 0.64 (n=158). Errors of omission and commission associated with predicted response reaches were 24.0% and 4.0%, respectively. As mentioned above, the use of the 150-meter arc sampling distance tended to minimize commission errors while increasing errors of omission between response and transport channel types. In theory these omission errors could be reduced by using a 100 meter arc sampling distance, but commission errors would likely increase. However, the ultimate limiting factor is the resolution and quality of the DEM data.

DISCUSSION

The intent of this effort was to produce a regionally consistent information base that federal agencies could use for planning or prioritizing salmonid habitat restoration opportunities in the PNW. Our analyses were based on simple concepts that are consistent with our understanding of habitat-forming processes in western Washington State. These are: (a) channel slope largely determines the range of potential channel morphologies; (b) large woody debris abundance modifies within channel type morphology; and (c) salmonid habitat utilization increases with increased

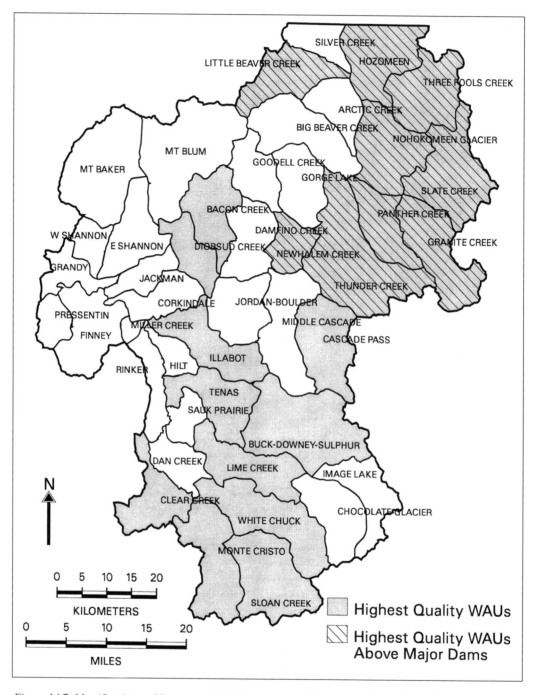

Figure 14.7. Identification and location of the highest quality WAUs in the Upper Skagit WRIA.

LWD abundance in the response reach channel type. We also presumed that large conifer riparian forests tend to be associated with greater LWD abundance than open or early seral stage riparian areas. Hence, the fundamentally important outputs of our analyses are the extent and location of response reaches (slope <0.04) and the condition of riparian forests along response reaches. The extent and location of response reaches identifies areas that may provide suitable habitat for

Table 14.5. Error Matrix Comparing Ground Visited Reference Data to the Predicted Stream Reach Data

Ground Visited Reference Data

	Response	Transport	Source	row total	% Correct	% Commission
Response	74	3	0	77	96	4
Transport	23	36	4	63	57	43
Source	0	3	15	18	83	17
column total	97	42	19	n=158		
% Omission	24	14	21			

Predicted Channel Types

Overall Accuracy
125/158 = 79%

$\hat{K} = 0.64$

salmonids, and riparian forest conditions indicate the likelihood that those reaches have the forced pool-riffle morphology that salmonids favor.

Our accuracy assessment generally supports the assumptions listed above. However, users of such data should be aware that, while the model typically underrepresents the extent of response reaches, areas identified as response reaches are likely to be correct. Field efforts designed to more accurately identify locations of potential salmonid habitat should therefore focus on areas identified as transport reaches. Field data suggest that those response reaches incorrectly identified as transport reaches are often located where tributary channels enter the valleys of larger channels.

The analyses were less accurate at predicting channel morphology within response reaches, although results generally support the hypothesis that increased forest age is associated with increased LWD abundance. Also, we found little difference in the proportion of forced pool-riffle channels between early and mid-forest seral stages. Histograms of GIS-generated data provide a broad-brush description of channel and riparian conditions at scales that are useful to managers with statewide or regional jurisdiction (Figure 14.4). These data provide a crude but comprehensive characterization of landscape and stream channel attributes that influence the abundance and condition of salmonid habitats.

A qualitative comparison between the preceding results and a field-based assessment of habitat losses in the Skagit River basin reveals that our GIS-based predictions are generally consistent with field data collected independently of this study. Based on the results of our analysis, we predict that the greatest habitat losses have occurred in the Skagit river floodplain and delta where little late seral stage forest remains. Beechie et al. (1994) found that by far the greatest proportion (73%) of coho salmon-rearing habitat losses were associated with diking, ditching, and dredging in the floodplain, and that these losses were associated primarily with urban and agricultural land uses. Hence, our GIS-based results at least grossly predict the same result as a field-based assessment.

Beechie et al. (1994) further noted that industrial forestry had less impact on coho-rearing habitat losses at the river basin scale, but was nevertheless strongly associated with habitat losses in tributary streams (channel widths <10 meters). Thus, forestry was associated with less severe impacts to coho salmon-rearing habitat than were urban and agricultural uses. Our results are also consistent with this relative ranking of severity of impact by type of land use. We show no late seral stage forest in WAUs where nonforest land uses dominated response reach zones, suggesting that the most severe impacts to habitat would be located in those WAUs. By contrast, we found a broader range of percent late seral stage in WAUs where forestry borders the majority of response reaches, indicating that impacts to rearing habitat should be less severe in those WAUs.

Although not all response reaches were bordered by late seral stage forest prior to European settlement, our results suggest a dramatic change in riparian conditions during the last 100 to 200 years. Prior to European settlement, forest fires, floods, and channel migration were dominant influences on stand ages and types near streams (e.g., Agee, 1988). Certainly these processes would create in a patchwork of stands along channel networks, resulting in a range of forest types and seral stages along response reaches. Our data for WAUs contained partially or fully within national parks and wilderness areas give some indication of this patchwork (Figure 14.8). The median percentage of response reaches in late seral stage WAUs located substantially in park and wilderness areas was 54%. This compares to 22% for commercial forestry, and <10% for urban-agriculture land uses. We caution, however, that the percentages shown in Figure 14.8A should not be construed as representative of "natural" conditions because many WAUs contain significant amounts of development.

In addition to a relative ranking, the data distributions can provide useful information for the development of preservation and/or restoration prescriptions. For example, some WRIAs have a

A

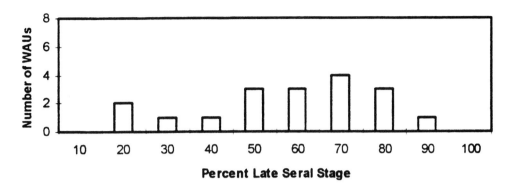

B

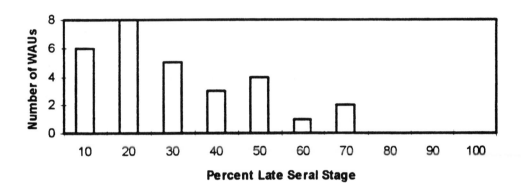

C

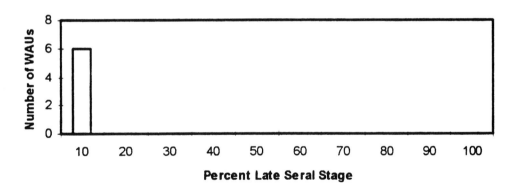

Figure 14.8. Frequency distribution of percent late seral stage along response reaches in WAUs dominated by (a) park and wilderness, (b) commercial forestry, and (c) urban-agriculture land uses.

relatively low percentage of response reaches in the late-seral forests, but a high percentage in mid-seral forests. A rational restoration consideration for these WRIAs may be the preservation of existing mid-seral forests in WAUs with a high density of response reaches. However, use of these analytical tools for identifying tasks or priorities for salmon habitat preservation and restoration can only be accomplished through a process that includes involvement of experts with knowledge of in situ habitat conditions. With the proper expertise and selected ancillary data (e.g., physical barriers to fish migration), map products identifying specific attributes of WRIAs and WAUs could provide a valuable data source to help prioritize the expenditure of preservation and restoration resources.

CONCLUSIONS

Our efforts demonstrate that remote sensing data and GIS methods can be applied to assess landscape attributes that influence the condition of salmon habitat at subbasin to watershed scales. GIS-based analytical products can be used to predict the locations of response reaches likely to provide salmon habitat. By using GIS buffering procedures along response reaches, the likelihood of finding a forced pool-riffle morphology based on the adjacent stream bank vegetation associations can be estimated. Both types of predictions have quantifiable error rates. These products could be used to target reaches where predictions are poor (e.g., the 23% of reaches predicted to be transport reaches that were response reaches), thereby increasing the efficiency of field efforts. Furthermore, such products can rapidly identify the quantity, extent, and condition of habitats at a scale useful for prioritizing regional protection or restoration efforts. We believe that such a wide-area, uniform database (uniform map themes and uniform coordinate system) can complement existing watershed screening protocols and help accomplish prioritization more rapidly and with greater reliability and objectivity.

SUMMARY

Categorization of 164,083 kilometers of stream length has provided the first quantitative measure of the extent and location of potential salmon stream habitat throughout western Washington State. Reach slope and forest seral stage provided a coarse indicator of channel condition across the region. Reach-average slopes calculated for individual stream reaches using 30-meter digital elevation model (DEM) data correctly identified low-gradient (<4.0% slope) response reaches that typically provide habitat for anadromous salmon with an accuracy of 96% (omission and commission error rates of 24.0 and 4.0%, respectively). Almost one-quarter (23.2%) of all stream length categorized consisted of response reaches, of which, only 8.7% were associated with late seral and 20.7% with mid-seral forest stages. Approximately 70% of the total stream length potentially providing anadromous salmon habitat is associated with nonforested and early-seral stage forests. GIS-based analytical techniques provided a rapid, objective, and cost-effective tool to assist in prioritizing locations of salmon habitat preservation and restoration efforts in the Pacific Northwest.

ACKNOWLEDGMENTS

The authors would like to acknowledge Bradford L. Johnson for graphics support. The U.S. Environmental Protection Agency (EPA) partially funded and collaborated in the research described here. It has been subject to the agency's programmatic review and has been approved for publication. Additional funding was provided by the Skagit System Cooperative, the U.S. Department of

Agriculture (USDA) Forest Service through Cooperative Agreement PNW–93–0441, and the USDA Cooperative State Research Service under Agreement No. 94–37101–0321. Mention of trade names or commercial products does not constitute endorsement or recommendation for use. Reproduced with permission, the American Society for Photogrammetry and Remote Sensing. Lunetta R., Cosentino B., Montgomery D., Beamer E. and Beechie T. "GIS-Based Evaluation of Salmon Habitat in the Pacific Northwest". *Photogrammetric Engineering and Remote Sensing*, Vol 63 no. 10 (October 1997), 1219–1229.

REFERENCES

Abbe, T.B., and D.R. Montgomery, 1996. Interaction of large woody debris, channel hydraulics and habitat formation in large rivers. *Regulated Rivers: Research and Management*, 12: 201–221.

Agee, J.K., 1988. Succession dynamics in forest riparian zones. In *Streamside Management: Riparian Wildlife and Forestry Interactions*. K.J. Raedeke, Ed., University of Washington, Institute of Forest Resources, Contribution No. 59.

Beechie, T. J., E. Beamer, B. Collins, and L. E. Benda, 1996. Restoration of habitat-forming processes in Pacific Northwest watersheds: A locally adaptable approach to salmonid habitat restoration. In *Role of Restoration in Ecological Management*, D.L. Peterson and C.V. Klimas, Eds. Society for Ecol. Restoration, Madison, WI, pp. 48–67.

Beechie, T., E. Beamer, and L. Wasserman, 1994. Estimating coho salmon rearing habitat and smolt production losses in a large river basin, and implications for restoration. *North American Journal of Fisheries Management*, 14: 797–811.

Bradbury, B. et al., 1996. Handbook for prioritizing native salmon and watershed protection and restoration. In review, 1996.

Collins, D., 1996. *The Rate of Timber Harvest in Washington State:* 1988–1991. Washington State Department of Natural Resources Publication, Forest Practices Division, Report No.1.

Congalton, R.G., R.G. Oderwald, and R.A. Mead, 1983. Assessing Landsat classification accuracy using discrete multivariate statistical techniques. *Photogrammetric Engineering and Remote Sensing*, 49(12): 1670–1678.

Congalton, R.G., and D.M. Schallert, 1992. Exploring the Effects of Vector to Raster and Raster to Vector Conversion. U.S. Environmental Protection Agency Internal Report, EPA/600/166.

Delong, M.D., and M.A. Brusven, 1991. Classification and spatial mapping of riparian habitat with applications toward management of streams impacted by nonpoint source pollution. *Environmental Management*, 15(4): 565–571.

Hudson, W.D., and C.W. Ramm, 1987. Correct formulation of the kappa coefficient of agreement. *Photogrammetric Engineering and Remote Sensing*, 53(4): 421–422.

Keller, E.A., and F.J. Swanson, 1979. Effects of large organic material on channel form and fluvial processes. *Earth Surface Processes*, 4: 361–380.

Lunetta, R.S., R.G. Congalton, L.K. Fenstermaker, J.R. Jensen, K.C. McGwire, and L.R. Tinney, 1991. Remote sensing and geographic information system data integration: Error sources and research issues. *Photogrammetric Engineering and Remote Sensing*, 57(6): 677–678.

Lunetta, R.S., J. Iiames, J. Knight, R.G. Congalton, and T.H. Mace, 2001. An assessment of reference data variability using a "virtual field reference database." *Photogrammetric Engineering and Remote Sensing,* 67(6): 707–715.

Montgomery, D.R., 1994. Geomorphological influences on salmon spawning distributions. Abstracts with Programs, *Proceedings of the Annual Geological Society of America Conference*, pp. A–439.

Montgomery, D.R., and J.M. Buffington, 1993. Channel Classification, Prediction of Channel Response, and Assessment of Channel Condition. Washington Department of Natural Resources Report, TFW-SH10–93–002.

Montgomery, D.R., and J.M. Buffington. Channel reach morphology in mountain drainage basins, 1997. *Geologicical Society of America Bulletin*, 109: 596–611.

Montgomery, D.R., J.M. Buffington, R.D. Smith, K.M. Schmidt, and G. Press, 1995. Pool spacing in forest channels. *Water Resources Research*, 31: 1097–1105.

Nehlsen, W., J.E. Williams, and J.A. Lichatowich, 1991. Pacific salmon at the crossroads: Stocks at risk from California, Oregon, Idaho, and Washington. *Fisheries*, 16(2): 4–21.

Oman, L., and L. Palensky, 1995. Preliminary Priority Watersheds for Restoration and Conservation of Fish and Wildlife. Washington Department of Fish and Wildlife Report, Olympia, WA.

Pacific Meridian Resources (PMR), 1993. Washington State Forest Cover and Classification and Cumulative Effects Screening for Wildlife and Hydrology. Final report submitted to Washington Department of Natural Resources, Olympia, WA.

Paustian, S.J., et al., 1992. A Channel Type User's Guide for the Tongass National Forest, Southeast Alaska. USDA Forest Service, Alaska Region 10, Technical Paper 26.

Peterson, N.P., A. Hendry, and T.P. Quinn, 1992. Assessment of Cumulative Effects on Salmonid Habitat: Some Suggested Parameters and Target Conditions. Washington Department of Natural Resources, Report No. TFW- F3–92–001.

Rosgen, D.L., 1994. A classification of natural rivers, *Catena*, 22(3): 169–199.

Story, M., and R.G. Congalton, 1986. Accuracy assessment: A user's perspective. *Photogrammetric Engineering and Remote Sensing*, 52(3): 397–399.

Swanson, F.J., and G.W. Lienkaemper, 1978. Physical Consequences of Large Organic Debris in Pacific Northwest Streams. U.S. Department of Agriculture Forest Service, Pacific Northwest Forest and Experiment Station, General Technical Report PNW–69, Portland, OR.

United States Department of Interior (USDI), 1993. Land Use and Land Cover from 1:250,000 and 1:100,000 Scale Maps. Geological Survey, Data Users Guide No. 4.

Washington Department of Natural Resources (WDNR), 1994. Data documentation for statewide classified canopy coverages.

Washington Department of Natural Resources (WDNR), 1993. Detailed Layer Description for Water Resource Inventory Areas. Forest Practices Division, Olympia, WA.

Washington Department of Natural Resources (WDNR), 1988. Detailed Layer Description for Watershed Administrative Units. Forest Practices Division, Olympia, WA.

CHAPTER 15

Physical Characterization of the Navarro Watershed for Hydrologic Simulation

Jeffrey D. Colby

INTRODUCTION

The Navarro watershed is an important natural resource in Costa Rica (Figure 15.1). The watershed is located at the headwaters of the Reventazón River basin, which may have the greatest hydropower potential of any drainage unit in the country (Quesada, 1979). High sediment loads and streamflow variability affect the proximate Cachí Reservoir (Quesada, 1979; Jansson and Rodriguez, 1992) and additional planned hydroelectric projects. Due to rich volcanic soils, the watershed is an important producer of agricultural products for the country (Cortés and Oconitrillo, 1987). In addition, groundwater is an important water supply source within the watershed.

The city of Cartago is located in the watershed and is part of the Gran Area Metropolitana (GAM). The GAM is the primary population center of the country, and increasing population growth raises important environmental issues (Monzón, 1993). For example, the effects of urban fringe growth on water resources is a concern of the Programa Nacional de Desarrollo Urbano Sostenible (PRODUS).

The 279 km² Navarro watershed is one of two located at the headwaters of the Reventazón River basin. The watershed exhibits diverse topography, and particularly dramatic relief (Figure 15.2). The elevation ranges from approximately 3300 m near the summit of the Irazú volcano to 1029 m at the La Troya streamflow gauge.

Management of surface and subsurface water in the Navarro watershed is important for hydroelectric power production and water supply maintenance. Simulating the hydrologic response of the watershed using spatially distributed characteristics could provide a valuable tool for hydrologic research and watershed management. This chapter will describe the physical characterization of the Navarro watershed using geographic information technologies and data to enhance distributed hydrologic modeling efforts. In particular, the processing of digital elevation data will be described.

The hydrologic model used to simulate runoff in the Navarro watershed was the Precipitation Runoff Modeling System (PRMS). The PRMS is a modular modeling system designed to evaluate impacts of climate and land use on surface runoff, sediment yields, and general basin hydrology (Leavesley et al., 1983). Parameterization of PRMS was undertaken based on modeling response units (MRU). MRUs are units of a watershed partitioned on the basis of characteristics such as vegetation type, precipitation distribution, slope, aspect, and soil type. The PRMS was operated within the Modular Modeling System (MMS). The MMS is an integrated computer software sys-

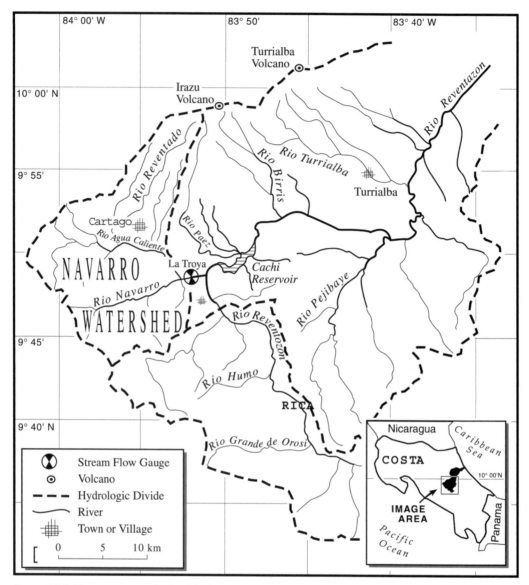

Figure 15.1. The upper watersheds of the Reventazón River Basin (Reprinted by permission, International Journal of Remote Sensing).

tem which provides an operational framework for development of algorithms and their application toward modeling physical processes (Leavesley et al., 1996).

PHYSICAL CHARACTERIZATION

The factors that influenced the selection of data for this study were requirements to complete the study and availability. In comparison to other Central American countries the amount and quality of environmental data available in Costa Rica is relatively high. However, initial assessment of data resources revealed significant gaps in some areas (e.g., soils data) and discrepancies in scales between others.

The availability of data for this watershed may be seen as deficient compared to that required

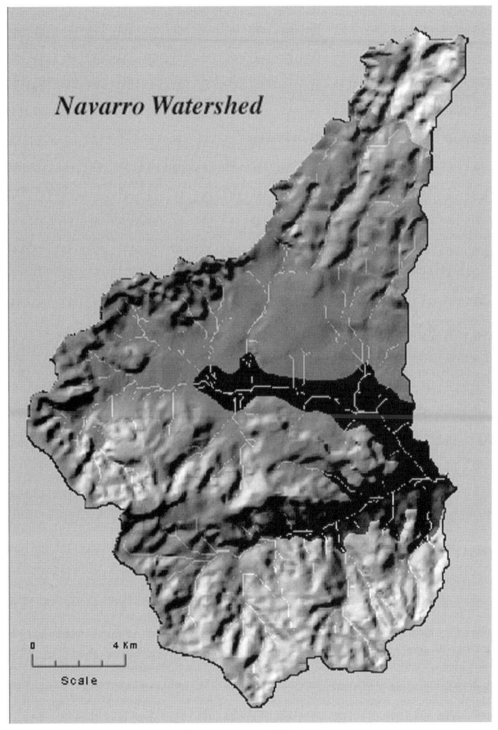

Figure 15.2. Shaded relief image of the Navarro Watershed.

for an experimental watershed in which it is desired to apply a fully distributed hydrologic model. However, this state of data availability lends itself to testing a distributed hydrologic modeling approach utilizing MRUs. In the MRU delineation process used in this study thematic data were aggregated, and homogenous units created from which more generalized parameter values were extracted for modeling purposes. Three types of data were utilized to develop GIS layers for the delineation of MRUs: elevation, digital satellite data, and precipitation.

Digital Terrain Elevation Data Processing

The primary elevation data used was the U.S. National Imagery and Mapping Agency (NIMA) Digital Elevation Terrain Data (DTED). Other elevation data considered bore constraints which were difficult to overcome. For example, topographic maps at 1:50,000 scale which covered the watershed were obtained, but digitizing the entire watershed was judged prohibitive for this study. In addition, complete coverage of the watershed by same scale stereo aerial photographs was not available for construction of a digital elevation model.

The DTED data were distributed in 1° by 1° cells, with a 16-bit range of elevation values. Two cells of DTED Level 1 data were utilized in this study. The boundaries of these cells extended from 9° N to 10° N and from 83° W to 85° W. According to the NIMA, the data for the cells were digitized or scanned from cartographic sources at a scale of 1:250,000, in the mid- to late 1970's. The DTED data were provided in a 3 arc second format. The resolution of this data between 9° N and 10° N latitude was 92.161 m^2. Elevation was represented by a regular grid of post points spaced at 100 m and interpolated with an inverse distance weighted routine using eight neighbors.

Preliminary Steps

A projection file was created in ARC/INFO to enable rectification of the original points from a geographic grid to a Lambert Conformal Conic projection. This projection was chosen because at low latitudes its conformality property possesses true shape of small areas, and also the ancillary maps available for the study area were represented in this projection. To enable integration with additional data sets the DTED file was resampled to 90 m using a bilinear interpolation routine.

Drainage Network and Watershed Delineation

A multiple step process was enacted in ARC/INFO GRID to define the drainage pattern and boundary of the watershed. Essentially four steps were carried out to delineate the drainage network using the elevation data: removing sinks in the DEM; assigning flow direction per cell; assigning flow accumulation values per cell; and determining the threshold flow accumulation value that best represented the drainage pattern (Jenson and Dominique, 1988).

In order to delineate the boundaries of the watershed, the drainage pattern was first displayed and the location of the streamflow gauge estimated. The streamflow gauge location on the derived drainage pattern was accurate in an east-west direction but approximately 550 m south of the actual location, according to a 1:50,000 scale topographic map. The topography in the area where the La Troya streamflow gauge was located opens to relatively level terrain, which may have contributed to the difficulty in defining the location of the gauge. The area draining to the streamflow gauge was identified using the Watershed function in GRID and the flow direction map. The resulting map defined the boundary of the watershed and provided a mask for delineating the boundary on additional GIS data layers.

Physical Characteristics Derived from Elevation Data

Additional analysis steps were carried out to derive physical characteristics of the watershed from the elevation data, including area, elevation parameters, perimeter of the watershed, and drainage density. These parameters were compared with parameters produced in previous studies of the watershed.

In a technical report generated by Solís et al. (1991) and a thesis by Baltodano and Hidalgo (1992), similar physical characteristics were extracted from 1:50,000 scale topographic maps for the Navarro watershed. In an earlier technical report, Elizondo (1979) extracted watershed characteristics from the 1:200,000 scale topographic map, San José CR2-CM–5, published by the Instituto Geographico Nacional (IGN). The report utilized the Puente Negro streamflow gauge which, due to flood damage, was later replaced downstream by the La Troya gauge.

In this study the area of the watershed was calculated using the GRID function Zonalarea. The perimeter was calculated using Zonalperimeter. The elevation parameters, such as maximum, minimum, and mean elevation values, were extracted in ERDAS IMAGINE. These parameters as well as drainage density were compared to estimates from Solis et al. (1991), Elizondo (1979), and estimates from the primary utility company in the country, the Instituto Costaricense de Electricidad (ICE) (Table 15.1).

Table 15.1. Physical Characteristics Derived from Elevation Data

Parameters Station	Colby L.Troya	Solis et al. L.Troya	Elizondo P.Negro	ICE L.Troya	ICE P. Negro
Area (km²)	278.57	282	273.3	273.6	273.3
Elevation max.	3427	3200	3300	—	—
Elevation min.	1057	1020	—	1028.6	1048.62
Elevation mean	1691	1725	1620	—	—
Perimeter (km)	108.54	87	78	—	—
Drainage density	0.66	0.88	0.46	—	—

Drainage density, or the length of streams per unit area, was calculated using (Black, 1991):

$$D_d = L/A \qquad (1)$$

where: D_d = drainage density
  L = length of streams (km)
  A = area of the watershed (km²)

Slope values for the watershed were computed in ERDAS IMAGINE and the GIS IDRISI. The following percentage distributions were calculated for the watershed:

(1). 0–4 degrees, 29.5%
(2). 5–9 degrees, 22.8%
(3). 10–15 degrees, 24.9%
(4). 16–19 degrees, 12%
(5). 20–28 degrees, 9.8%
(6). 29–48 degrees, 1%

Elevation Data Quality Assessment

An advantage of using DTED data in this study was the ease with which parameters such as physical characteristics were derived. Use of previously derived digital elevation data provided flexibility and analysis capabilities in a timely fashion. The trade-off was the lack of quality control in data development.

The drainage pattern derived from the DTED data provided a generally representative depiction of the watershed. However, errors were encountered. Accuracy was assessed by comparisons to drainage patterns from a 1986 Landsat Thematic Mapper (TM) image, 1:10,000 scale land use maps, 1:50,000 scale topographic maps, and aerial photographs. An overlay of the drainage pattern on a false color composite TM image (Figure 15.3, see color section) revealed nonsystematic errors in the correspondence of the location of the drainage pattern, especially along the slopes of the Irazú volcano in the northern section of the watershed. In other areas the drainage patterns matched.

These errors were due in part to the quality of the DTED data and the level of sophistication of GIS software tools used for drainage pattern extraction. The poor representation of the rivers in the northern section of the watershed raised the question of whether the cartographic source for the elevation values was produced before the 1963–1965 eruption of the Irazú volcano.

The most recent eruption of the Irazú volcano began in March of 1963 and continued to spew large volumes of lithic ash through March of 1965. Accumulation of ash on the slopes of the volcano altered the hydrologic regime of the rivers. A hard impervious crust formed on the mantle of ash which reduced infiltration, increased runoff, and resulted in increased slope erosion, frequent flash floods and deadly debris flows. Emergency measures were undertaken such as channel improvements, the construction of levees to protect Cartago, and watershed rehabilitation by terracing, drainage diversion, contour trenching and artificial revegetation (Waldron, 1967).

The area, mean elevation, and drainage density values derived for the Navarro watershed fell between the values produced from analog sources (Table 15.1). The maximum and minimum elevation and perimeter values calculated from the DTED data were somewhat higher than that produced from the 1:50,000 and 1:200,000 scale topographic maps. The absolute vertical accuracy of the DTED data may have affected the differences in elevation values, and the perimeter differences were likely due to grid cells representing the watershed outline, rather than a vector line. Overall, the quality of the elevation data and the derived drainage pattern, though not optimal, supported its application in a distributed hydrologic modeling exercise utilizing MRUs.

Land Cover Classification

The digital Landsat TM satellite image was classified to provide land cover characterization. Variable illumination angles and reflection geometry due to different slope and aspect orientations limit the effectiveness of Landsat TM classification efforts in mountainous terrain. A method for reducing the resulting anisotropic reflectance effects, which was developed in a temperate region (Smith et al., 1980; Colby, 1991; Hodgson and Shelley, 1994) was tested in the neotropical environment of central Costa Rica (Colby and Keating, 1998).

Aerial photographs and land use maps provided reference information for classification of the TM image. Black and white aerial photographs for the north-central section of the watershed (1989) at a scale of 1:20,000 and for the southern section of the watershed (1992) at a scale of 1:60,000 were obtained from the IGN. Also, 1:10,000 scale land use maps (field checked in 1989) were obtained for the central part of the watershed.

Improved classification accuracy was obtained using topographic normalization routines (Colby and Keating, 1998). The resolution and quality of the elevation data was believed to have reduced topographic normalization effectiveness and classification accuracy.

Precipitation Distribution

The Navarro watershed is located between diverse precipitation regimes. One of the driest areas in the country extends from Cartago through the Valle Central (Janzen, 1983). The highest rainfall rates in the Upper Reventazón River basin, 7500 mm a year, are found approximately 25 km southeast of Cartago (Jansson and Rodriguez, 1992). The mean annual precipitation recorded at stations used in this study varied from 1329 mm at the centrally located Comandancia station to 3264 mm at the Belen station located in the southeast near the southern border of the watershed.

To develop an image of the spatial distribution of precipitation across the watershed, a number of trend surface analyses were undertaken using the GIS IDRISI. Second and third order trend surfaces were derived using mean annual rainfall figures calculated from complete years of data for the 20 year period between 1967 through 1986 (IMN, 1988). Trend surfaces were created using data from 7 to 22 stations located in and near the watershed.

Generally, third order surfaces had the highest goodness of fit (R^2) values, but did not represent the spatial distribution of precipitation in the southwest section of the watershed accurately due to a scarcity of precipitation stations in the area. The final image chosen was a second order trend surface created using 14 stations. The R^2 value for this surface was 91.35%.

APPLICATIONS

Once digital spatial characteristics of the watershed had been generated MRUs were delineated using thematic GIS data layers of watershed subbasins, a distance buffer from the stream, precipitation distribution, and land cover categories. The subbasins were delineated using the same techniques as described above to delineate the watershed; however, research and management criteria were taken into consideration in determining outflow points. A distance buffer from the stream was created based roughly on the variable source area concept (Troendle, 1985). The precipitation distribution layer consisted of a three category aggregation of the second order trend surface described above. The land cover layer included the following categories: bare areas, grass, shrubs, trees, and impervious areas. MRUs were delineated using the four thematic layers and a GIS-based methodology (Colby, 1995).

Following delineation of the MRUs, hydrologic simulation of the watershed was undertaken for August through January during the 1987–1988 Costa Rican water year. Parameters for PRMS were extracted using the MRU boundaries, the MRU thematic layers, and additional data layers such as elevation, slope, and aspect. The accuracy of the simulations was determined to be sufficient to proceed with a scaling analysis using multiple resolutions of land cover data.

In the scaling analysis a series of MRU patterns aggregated at 90 m^2 intervals, from 90 m^2 to 1260 m^2, provided areal dimensions to parameterize PRMS. Simulation of hydrologic runoff was undertaken at each resolution. The fractal dimension D values of the MRU patterns were also calculated at each resolution. A strong correspondence was found between the range of resolutions at which accurate hydrologic simulations were achieved and the range of self-similarity of the MRU patterns, as measured by their fractal dimension (Colby, 2001).

DISCUSSION AND CONCLUSION

The data available for this study were sufficient to carry out the intended applications; however, shortcomings did exist. An obvious gap was the lack of soils data available at a useful resolution. The resolution and quality of the elevation data affected the derivation of several watershed characteristics and anisotropic reflectance correction efforts. The status of data available for this study may be representative of many watersheds in tropical countries in which watershed modeling and management are desired. For example, in these countries topographic maps at a scale larger than 1:200,000 or 1:250,000 may not be available. Physicial characterization of this watershed provided the opportunity to evaluate the effectiveness of utilizing a distributed hydrologic modeling approach based on MRUs. The capability to aggregate thematic data and form hydrologic units provided flexibility to work with less than optimal data. The MRU modeling approach delivered effective watershed characterization and hydrologic simulation accuracy which enabled the desired research to be accomplished. Utilization of geographic information systems provided essential capabilities for improving the processing, management, and analysis of available data.

ACKNOWLEDGMENTS

The author would like to acknowledge the financial support of the Organization of American States, along with the assistance of the U.S. Geological Survey, the U.S. Defense Mapping Agency, and numerous organizations and individuals in Costa Rica for their support in this research effort.

REFERENCES

Baltodano P., J.A., and H. Hidalgo L., 1992. *Definición de Niveles de Inundación en el Río Reventado*. Proyecto de graduacion, Escuela de Ingeniercia Civil, Universidad de Costa Rica, San José, Costa Rica.

Black, P.E., 1991. *Watershed Hydrology*. Prentice-Hall Inc., Englewood Cliffs, NJ.

Colby, J.D., 1991. Topographic normalization in rugged terrain. *Photogrammetric Engineering and Remote Sensing*, 57(5):531–537.

Colby, J.D., 1995. Resolution, Fractal Characterization and the Simulated Hydrologic Response of a Costa Rican Watershed. Ph.D. Dissertation, Geography, University of Colorado, Boulder, CO.

Colby, J.D., 2001. Simulation of Costa Rican Watershed: Resolution Effects and Fractals. *Journal of Water Resources Planning and Management*. 127(4).

Colby, J.D., and Keating, P.L., 1998. Land cover classification using Landsat TM imagery in the tropical highlands: The influence of anisotropic reflectance. *International Journal of Remote Sensing*, 19(8), 1479–1500.

Cortés G., V.M., and G. Oconitrillo C., 1987. Erosion de Suelos Horticulas en el Area de Cot y Tierra Blanca de Cartago. Tesis de Grado para Optar al Grado de Licenciado en Geografia. Departmento de Geografia, Universidad de Costa Rica, San José, Costa Rica.

Elizondo M., J.A., 1979. Estudio Hidrogeológico Preliminar de la Cuenca del Rio Navarro Provincia de Cartago. Servicio Nacional de Aguas Subterraneas Riego y Avenamiento, Departmento de Ingeniería e Hidrología, San José, Costa Rica.

Hodgson, M.E., and B.M. Shelley, 1994. Removing the topographic effect in remotely sensed imagery. *ERDAS Monitor*. 6(1)4–6.

Instituto Meteorológico Nacional (IMN), 1988. Catastro de las Series de Precipitaciones medidas en Costa Rica. San José, Costa Rica.

Jansson, M.B., and A. Rodriguez, Eds., 1992. Sedimentological Studies in the Cachí Reservoir, Costa Rica. Department of Physical Geography, Uppsala University, UNGI Rapport no. 81 Uppsala, Sweden.

Janzen., D.H., 1983. *Costa Rican Natural History*. The University of Chicago Press, Chicago, IL.

Jenson, S., and J. Dominique, 1988. Extracting topographic structure from digital data for geographic information system analysis. *Photogrammetric Engineering and Remote Sensing*, 54(11):1593–1600.

Leavesley, G.H., R.W. Lichty, B.M. Troutman, and L.G. Saindon, 1983. Precipitation-runoff modeling system — User's manual: U.S. Geological Survey. Water Resources Investigations Report, 83–4238.

Leavesley, G.H., P.J, Restrepo, L.G. Stannard, L.A. Frankoski, and A.M. Sautins, 1996. MMS: A Modeling Framework for Multidisciplinary Research and Operational Applications. In *GIS and Environmental Modeling: Progress and Research Issues,* M.F. Goodchild, L.T. Steyart, and B.O. Parks, Eds., GIS World Books, pp. 155–158.

Monzón P., J.P., 1993. Patrones de Crecimiento del Gran Area Metropolitano. Tesis de Grado para Optar al Grado del Licenciado en Ingeniería Civil, Escuela de Ingeniería Civil, Universidad de Costa Rica San José, Costa Rica.

Quesada M., C.A., 1979. Effect of Reservoir Sedimentation and Streamflow Modification on Firm Power Generation. Ph.D. Dissertation, Department of Civil Engineering, Colorado State University, Fort Collins, CO.

Smith, J.A., T.L. Lin, and K.J. Ranson, 1980. The Lambertian assumption and Landsat data. *Photogrammetric Engineering and Remote Sensing*, 46(9):1183–1189.

Solis B. H., M. W., Murillo and R. Oreamuno V., 1991. Estudio Hidrologico e Hidráulico para el Control de Inundaciones en la Cuenca del Rio Purires: Valle de Guarco. Servicio Nacional de Aguas Subterraneas Riego y Avenamiento Dirección de Ingeníera, Centro Agronomico Tropical de Investigación y Enseñanza: Proyecto Renarm-Cuencas, Costa Rica.

Troendle, C.A., 1985. Variable source area models. In *Hydrological Forecasting*, M. Anderson, and T. Burt, Eds., John Wiley and Sons, New York, NY, pp. 347–403.

Waldron, H.H., 1967. Debris Flow and Erosion Control Problems Caused by the Ash Eruptions of Irazú Volcano, Costa Rica. Geological Survey Bulletin 1241–I. United States Government Printing Office, Washington, D.C.

CHAPTER 16

Hydrologic Modeling Using Remotely Sensed Databases

James F. Cruise and Richard L. Miller

INTRODUCTION

Recent developments in the evolution of hydrologic modeling using remotely sensed data sources are discussed. The latest remote sensing instruments as well as traditional devices are addressed in terms of spectral characteristics and spatial resolutions. Products of interest to hydrologic modelers are described, as are techniques for their derivation. Hydrologic modeling techniques that make use of these data sources are examined, particularly at the macroscale level. Scientific development in the areas of global change analysis and atmospheric forecasting establish the context of research in hydrologic modeling using spatial databases. Of primary importance are issues involving data scale and resolution requirements for hydrologic modeling. The evolution of this research is discussed and the authors present a study that makes use of recently developed modeling techniques along with remote sensing data products. The results demonstrate that macroscale hydrologic modeling in conjunction with conventional remote sensing data sources can lead to accurate simulations of observed climatologies over fairly long time periods.

Background

Spatial databases are frequently a critical component of hydrologic models. The requirements for modeling over larger spatial domains imposed by the global change and atmospheric forecasting communities have dramatically altered some conceptions relative to modeling strategies and data requirements. The formulation of coupled atmospheric/hydrologic modeling systems to operate in the macroscale spatial domain has placed a premium on data acquisition and processing techniques. The development of comprehensive databases, and hence effective models, is therefore a major focus of most modeling efforts. Remote sensing can provide unique data for constructing spatial databases. These data may include analog aerial photography or digital images representing the emission or reflectance of radiant energy from ground surfaces. Remotely sensed data are available from a variety of instruments mounted on satellite and aircraft platforms. Following acquisition, these data must be processed to correct for the influence of the intervening atmosphere, remapped to a geographic or spatial grid, and analyzed to derive information required specifically for hydrologic modeling.

The purpose of this chapter is to present some of the latest developments in the use of remotely sensed data and geographic information system tools in hydrologic modeling. Sources of remotely sensed data, capabilities of important sensors, acquisition techniques, and hydrologic modeling strategies to make use of these data will be discussed. Data requirements relative to temporal and spatial scale of modeling domains will be examined, primarily within the context of global change hydrology. Finally, a recent study completed by the authors that made use of many of the data sources and modeling techniques discussed in the chapter will be presented.

DATA REQUIREMENTS FOR HYDROLOGIC MODELING

Hydrologic models require different types of data depending on the processes modeled and the relevant time and space scales of these processes. Models are generally classified temporally as either single event or continuous. Single event models are designed to simulate a single climatic event and thus do not attempt to model hydrologic conditions between events. Continuous models attempt to simulate all hydrologic conditions on a watershed at some specified time step (e.g., hourly or daily) for a long period of time (months or years). In general, continuous simulation models require more extensive climatic data and, possibly, soils and vegetation data sets than do single event models.

Different modeling strategies describe the spatial variation of watershed characteristics at various levels of detail and complexity. Models are classified spatially as either lumped or distributed. The least complex are the lumped parameter models where the spatial variability of select watershed characteristics are described by a few simple parameters (e.g., *lumped* together). Typical parameters include watershed lag (time from centroid of precipitation excess to centroid of runoff hydrograph), time of concentration (travel time from most remote point in the watershed to the basin outfall), and storage factor.

The unit hydrograph method is an outstanding example of a lumped parameter hydrologic model. The more spatially complex distributed models attempt to describe more fully the spatial variation in topography, soils, surface characteristics and meteorology and to explicitly include these in the model. When using this method, the watershed surface is discretized into a spatial grid and the characteristics of the watershed are described within each grid cell. Between the extremes of the lumped and distributed models, several semidistributed techniques have recently been developed which combine the advantages of the distributed models with the simplicity of the lumped parameter method. Of these, the hydrologic response unit technique is quite popular and will be described in further detail in the applications section of this chapter.

All hydrologic models require physical data such as topography and soils characteristics, land-use data such as land cover and vegetation characteristics, and climatic data such as precipitation and temperature. In addition, data quantifying some aspect of streamflow within the watershed are generally required for model calibration and verification.

In constructing databases for hydrologic modeling, consideration must be given to the data requirements for model calibration and the spatial and temporal scale of the model. Products of interest to hydrologic modelers that can be derived from spatial databases include surface representation (slope, aspect, plain geometry), soils characterization (permeability, conductivity, storativity), antecedent moisture conditions, land cover and condition, and vegetation characteristics and biomass. Input parameters are either included directly in the database (i.e., hydraulic conductivity) or calculated from the raw data layers (i.e., slope, land cover). Traditionally, distributed models have employed databases at spatial resolutions on the order of meters. The models may execute on a time scale of minutes or hours, but normally the spatial database is updated on time scales of months or seasons to take into account land use and vegetation changes.

Unfortunately, it is not feasible to construct databases at optimal spatial resolutions for hydrologic modeling on anything but the smallest basin or subbasin scale due to the lack of suitable data and computational resources for the model. Zhang and Montgomery (1994) have suggested a grid resolution of 10 m as appropriate for hydrologic modeling. However, at this resolution the data are usually not available over large areas. Hydrologic studies are normally conducted using raster (gridded) data sets at spatial resolutions of 30 m or less if possible. More recently, semi-distributed macroscale models constructed at coarser resolutions (1–4 km) have been developed, particularly to function with atmospheric circulation models (Kite et al., 1994; Nijssen et al., 1997).

A number of researchers have conducted investigations of the effect of scale and data resolution on the accuracy of hydrologic models. Wolock and Price (1994) used a topographically driven hydrologic model (TOPMODEL) to investigate the effects of map scale and resolution on hydrologic modeling. They compared results using the 1:24,000 scale USGS maps at grid resolutions of 30 m and 90 m and the 1:250,000 scale, 90 m resolution maps. In general, they found that runoff estimates tended to increase (as did the ratio of surface runoff to total runoff) with increasing map scale and resolution. They attributed this observation to the effect of increasing grid size on the topographical index in the model. However, they did not conclude that the finer scale and resolution maps were necessarily better for hydrologic modeling purposes. Instead, they pointed out that the surface of the saturated soil water zone (which controls runoff generation) may be smoother than fine-scale surface topography, and so would correspond better to coarser resolution data. Likewise, Zhang and Montgomery (1994) found that runoff tended to increase with increasing grid cell sizes in a study based upon topographical information obtained from a 1:4,800 scale base map. In this study grid cell resolutions varied from 2 to 90 m. The authors found that slope representations of an area decreased with increasing grid size, while contributing drainage area increased with grid size. On a similar point, Wood et al. (1988) reported that runoff from areas of variable topography and soils characteristics demonstrated minimum variance at a resolution of about 1 km^2, thus indicating that hydrologic modeling with coarse raster data sets is reasonable. The macroscale model results reported by Kite et al. (1994) and Nijssen et al. (1997) also appeared to show that coarse resolution modeling holds considerable promise.

Representation of basin topography and soil characteristics can be accomplished using either vector- or raster-based data sets. Watershed surface topography is represented by use of a Digital Terrain Model (DTM). The DTM represents the topographic characteristics of the basin through either a grid cell (raster) or data string (vector) formulation (Sole and Valanzano, 1996). The raster-based digital elevation model (DEM) data sets available on-line provide elevation data at grid sizes ranging from 90 to 200 m. Vector-based digital line graph (DLG) data are essentially digitized versions of contour strings from standard topographic maps. The contour intervals range from 5–10 ft (1.5 to 3.0 m) to 50 m.

Surface features may be represented within the GIS by using the raw topographic data in raster or vector format, or by refinements such as the Triangular Irregular Network (TIN). In this method, an area is represented by irregular triangles with the elevation specified at each triangle vertex. The elevation is assumed to vary linearly over the triangle (Singh and Fiorentino, 1996). This formulation allows for the specification of irregular features which might not be possible with gridded data and is commonly employed in hydrologic modeling (Silfer et al., 1987; Maidment et al., 1989; Goodrich et al., 1991; Greene and Cruise, 1996).

Land cover data for hydrologic modeling purposes are usually obtained from interpretation of aerial photographs or multispectral data acquired by satellite or aircraft mounted sensors. High resolution black and white or infrared aerial photographs are available for most of the United States through the U.S. Geological Survey. The photographs are available at scales of 1:30,000 or less for most places. The Survey periodically determines land cover for the U.S. by interpretation of these photographs. Multispectral data sources and products derived from them are discussed later in this chapter.

ON-LINE DATA SOURCES

Several large databases that are available on the internet may be valuable for hydrologic modeling, particular at the macroscale level. In general, these databases are maintained at low spatial resolutions. A valuable source of hydrologic, topographic and land cover data for the United States is the EROS Data Center (EDC) of the U.S. Geological Survey. Topographic data are archived at the EDC in both Digital Elevation Model (DEM) and Digital Line Graph (DLG) format, depending on the location. For instance, the largest scale topographic maps that are commercially available for the U.S. are the 7.5′, 1:24000 scale maps (30m resolution) produced by the U. S. Geological Survey. However, these maps have only been digitized for select areas of the country. Instead, the 15′, 1:250,000 scale digital elevation model (DEM) maps are universally available at the EDC for topography, while the 1:100,000 scale digital line graph maps provide hydrography and transportation data. The grid cell size of the raster-based 15′ DEM data varies from 90 m to 200 m. This type of coarse resolution data may be appropriate for macroscale hydrologic modeling. The EDC is also the repository of the Land Use and Land Cover (LULC) database that is composed of 200 m raster images derived from aerial photographs that cover the entire conterminous United States.

The availability of fine resolution digital soils data for the U. S. is more problematic than are the topogaphic and hydrographic maps. The highest resolution soils data commonly employed in the U.S. are the county soil surveys conducted by the National Resources Conservation Service (NRCS). These surveys are mapped onto aerial photographs at a scale of 1:12,000 or 1:63,360 with individual soil series as the mapping units. NRCS provides soils data in both raster and vector formats at various scales and spatial resolutions. The Soil Survey Geographic Data Base (SSURGO) consists of the digitized county soils maps in vector (DLG) format. These files are released as they are completed and verified and are currently available only in scattered areas of the country.

The State Soils Geographic Database (STATSGO) is a generalization of the county surveys in which the individual mapping units have been aggregated based upon statistical analysis of the soil series. STATSGO is a vector database in which the smallest mapping unit is 1546 acres. The associated data tables include most of the soil matrix and parameter information contained in hard copy surveys. The Penn State University Environmental Sciences Systems Center has developed the Conterminous United States Soils (CONUS) database. CONUS is a raster database at 1 km resolution which represents the STATSGO soils polygons in a standardized 11 layer matrix format.

For land-use determination, various satellite mounted sensors are available that acquire multispectral data at spatial resolutions from 10 m to 1.1 km and temporal resolutions from 6 hours to 16 days. These data sets are used to obtain images of an area of the earth's surface that can be classified in terms of land cover or vegetation characteristics. Classified images are routinely used to obtain land cover information needed in hydrologic modeling such as forest canopy cover and/or biomass, agricultural crop coverage and characteristics (cultivated or noncultivated), urban land use acreages, etc. (Tan and Shih, 1988; Wilkening, 1989; Kite and Kouwen, 1992; Miller and Cruise, 1995). Satellite data are archived in a number of institutions in the United States. The Distributed Active Archive Centers (DAAC) associated with the NASA Earth Observing System Data and Information System (EOSDIS) network are a primary source of satellite data. The EDC serves as the DAAC for the global 1.1 km resolution images from the Advanced Very High Resolution Radiometer (AVHRR) instrument. These images have been archived since 1974. The Marshall Space Flight Center (MSFC) is the primary archive for data associated with the hydrologic cycle.

REMOTE SENSING

A useful operational definition of remote sensing is *to observe an object at some distance without contacting the object.* The large observational scale and unobtrusive nature of remote sensing make this technology a valuable tool for the study of multiscale features of the earth's surface. Although remote sensing cannot provide direct measurements of most hydrologic processes, remote sensing and related technologies such as image processing and Geographic Information Systems (GIS) can provide critical data and products for parameterizing many hydrologic models.

Remotely sensed data are obtained using several different data acquisition technologies. In general, images are formed based on the reflected or emitted energy from a defined spatial element (e.g., pixel, or ground resolution) over select interval(s) of the electromagnetic spectrum. A remote sensing instrument measures the energy at the detectors from either passive (ambient) or active (instrument-generated) energy sources. Although nonimaging instruments such as satellite altimeters do exist, this discussion will focus on the principles and data products derived from imaging instruments, particularly digital multispectral (*few* spectral bands) and hyperspectral (*many* spectral bands) sensors which provide useful data for application in hydrologic modeling. For example, standard CIR (color infrared) aerial photographs, or images, are formed by exposing film to reflected visible and infrared energy. This is an analog process. Important information on landscape patterns, land-use characteristics, and ground cover vegetation can be extracted from such aerial photographs using basic photogrammetric techniques (see for example, Paine, 1981; Graham, and Read, 1986). However, the processing and analysis of photographs using these techniques are time-consuming and labor-intensive. In contrast, the recent rapid development of digital multi- to hyperspectral instruments and low-cost computer systems has established digital remote sensing as an easy-to-use, affordable, and efficient tool for monitoring earth system processes (Miller, 1993; Miller and DeCampo, 1994; Miller et al., 1995).

Multispectral classification is a primary tool for translating remotely sensed data for use in many basic and applied research disciplines. This method is used here to provide a comprehensive example of the processes involved in applying remote sensing to hydrologic models. Multispectral classification is based on a spatial analysis and grouping of surfaces with a similar or correlated reflectance spectrum (i.e., spectral signature). Figure 16.1 shows representative spectra for several surface types along with the spectral bandpass (detection intervals) for the Landsat Thematic Mapper (TM), a satellite-based sensor. A ground point or image pixel is assigned to an image *class* if the pixel's spectral signature is statistically similar to the class mean spectrum. Statistical criteria are derived using multivariate classification methods such as the maximum likelihood classifier, principal component analysis, generalized eigenvalue analysis, and artificial neural networks (Jensen, 1986; Lillesand and Kiefer, 1993; Smith et al., 1994; Venkateswarlu and Singh, 1995; Roger, 1996; Atkinson and Tatnal, 1997; Murai and Omatu, 1997; Paola and Schowengerdt, 1997) These methods yield classified images of land-cover/land-use, soil type and moisture, landscape or drainage patterns, and hydrologic units (Bober et al., 1996; Cialella et al.,1997; Homer et al., 1997; Su et al., 1997; Van Deventer et al., 1997). The digital format of these products allows for a greatly simplified numerical analysis and data extraction process through the use of readily available image processing software. Classification results can then be directly integrated into a GIS, a hydrologic model, or used to derive model coefficients.

The utility of a classified image in hydrologic modeling depends on the accuracy of the classification scheme employed. Classification accuracy is related to the variability in the reflectance spectra for each derived class, band position and width, radiometric sensitivity and noise (e.g., spectral characteristics) of the instrument, and numerical precision of the classification algorithm.

Both the spectral shape and reflectance of a class can vary significantly in time and space as a

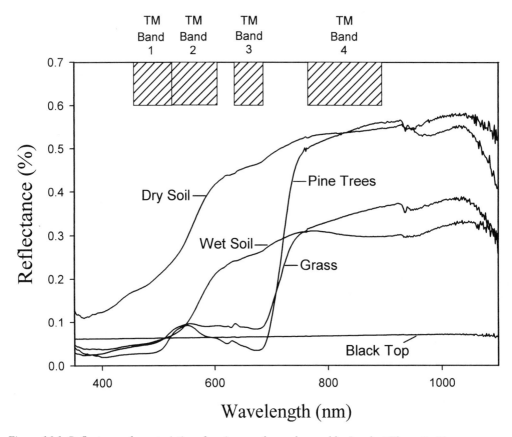

Figure 16.1. Reflectance characteristics of various surfaces observed by Landsat Thematic Mapper.

complex function of many processes. For example, the percent reflectance of vegetation at several wavelengths will increase with decreasing chlorophyll pigment concentration in response to physiological stress (Carter, 1993; Carter and Miller, 1994; Carter et al., 1996). The reflectance of soils in the visible spectrum will change as a function of soil moisture. Surface soil moisture can be measured using passive reflectance and microwave instruments through detecting changes in the soils physical and dialectric properties (Engman, 1990; Jackson et al., 1996; O'Neill et al., 1996). Synthetic Aperture Radar (SAR) systems are also useful in measuring soil moisture (Poncet et al., 1994; Sabburg, 1994; Boisvert et al., 1997; Wegmuller, 1997). The classification of pervious or impervious urban surfaces also changes with surface moisture and temperature. Hence, the ability to detect and resolve changes resulting from intraclass variability from a transition to a different hydrologic unit is an important aspect of current remote sensing research.

Recent advances in remote sensing technology and algorithm development are reducing misclassifications and improving the definition of the hydrologic response of various hydrologic units. Multispectral instruments are generally designed for a specific application. The spectral bands available may be less than optimal for other applications, resulting in large classification errors. An increase in the number of an instrument's spectral bands, as well as narrower band widths, will provide more accurate estimates of the shape (e.g., inflection points) of the surface reflectance spectrum for each class. Narrower bands also help isolate changes in reflectance to specific physical, chemical, and biological processes. Hyperspectral instruments (e.g., AVIRIS, CASI,

Table 16.1. Characteristics of Remote Sensing Instruments

Sensor	Number of Bands	Spectra Range (μm)	Band Width (μm)	Major Products
Advanced Visible/Infrared Imaging Spectrometer (AVIRIS)	224	0.40–2.50	~0.01	1247
Airborne Terrestrial Applications Scanner (ATLAS)	15	0.45–1.75 2.08–4.20 8.20–12.2	~0.1 0.27–0.85 0.40–1.0	
Calibrated Airborne Multispectral Scanner (CAMS)	9	0.45–12.50 2.08–2.35 10.40–12.50	0.06–0.2 (Vis./Near IR) 0.28 (Mid. IR) 2.1 (Thermal IR)	123467
Compact Airborne Spectrograpic Imager (CASI)	288	0.40–1.00	~ 0.002 (adjustable)	12347
European Remote-Sensing Satellite, Microwave Sounder (ERS-1 & ERS-2)	2	23.8 and 36.5 GHz	400 MHz (each band)	4,5
European Remote-Sensing Satellite, Active Microwave Instrument Synthetic Aperture Radar (ERS–1 & ERS–2, AMI-SAR)	C-Band	C-Band (5.3 GHz)	15.55 MHz	45
Hyperspectral Digital Imagery Collection Experiment (HYDICE)	210	0.40–2.50	~ 0.01	1247
Japanese Earth Resources Satellite, Synthetic Aperture Radar (JERS-1 SAR)	L-Band	1.2575 GHz	15 MHz	4
Multispectral Scanner (MSS)	4	0.50–1.10	0.10–0.31	1247
Système Probatoire d'Observation de la Terre (SPOT)	3 1 (Pan)	0.50–0.89 0.50–0.73	~ 0.1 0.22	127
Thematic Mapper (TM)	7	0.45–1.75 2.08–2.35 10.40–12.50	0.06–0.2 (Vis./Near IR) 0.28 (Mid. IR) 2.1 (Thermal IR)	12347
Thermal Infrared Multispectral Scanner (TIMS)	6	8.20–12.20	0.04–0.1	36

1. Land-cover / land-use.
2. Hydrologic units.
3. Surface temperature (thermal analysis).
4. Soils (soil moisture).
5. Precipitation
6. Evapotranspiration.
7. Runoff analysis

HYDICE) with hundreds of bands are effectively imaging spectrometers and may be used for a broad range of applications. The user simply chooses the bands most appropriate for a particular application based on a priori information or comprehensive processing to determine which may reveal information most relevant to a given study. Table 16.1 lists the characteristics of current and planned remote sensing instruments and examples of derived products for hydrologic modeling.

Algorithms based on a combination of bands reduce the effects of nonclass processes. For example, data from two bands, the red and near infrared, are used to compute the Normalized Difference Vegetation Index (NDVI). The NDVI is a simple, yet robust, algorithm to estimate vegetation abundance or *greenness* (Jensen, 1986; Goetz, 1997; Rasmussen, 1997; Ricotta et al., 1997). Similarly, hyperspectral sensors provide the potential to develop highly refined algorithms for classifying ground surfaces and hydrologic processes. Additional research in the spectral response of class components (e.g., plants, soils) to a range of processes will significantly improve classifications by providing more reliable training fields in supervised (interactive) classification algorithms using hyperspectral data.

Remote sensing technology is well established for measuring hydrologic variables or processes related to transport, snow and ice reservoirs, evaporation, and precipitation. The movement of water through rivers, estuaries, and coastal waters can be monitored using reflectance and thermal imagery. For example, a time series of red and near infrared reflectance images may indicate water movement using suspended particulates as a tracer (Miller and Cruise, 1994; Miller et al., 1994, Schultz, 1996). Multiple images of thermal data can indicate water mass structure, mixing, and flow patterns between areas of different temperatures. Thermal imagery is also used to measure evapotranspiration of forested landscapes (Luvall and Hobo, 1991; Courault et al., 1996) and evaporation in urban environments (Lo et al., 1997). The integration of data from multiple sources and remote sensing can help develop comprehensive models for analyzing the hydrology of a watershed, associated coastal zone, or components of the hydrologic cycle.

CASE STUDY: RIO GUANAJIBO, PUERTO RICO

In this section, we describe a recent study completed by the authors that made use of many of the data sources and hydrologic methods described in the previous sections. A joint project between the NASA Science and Technology Laboratory at Stennis Space Center, MS and the University of Puerto Rico was initiated in 1987 to examine the effects of changing landscapes on the water quality of local rivers and Mayaguez Bay, Puerto Rico (Otero et al., 1992). A particular focus of the study was to examine the correlation of anthropogenic activities such as agricultural practices, deforestation, and urbanization to increased sediment and nutrient-enhanced productivity in the bay. As part of this effort, a mathematical model was used to simulate runoff and sediment yield from the contiguous drainage basins over various periods of time (Cruise and Miller, 1993, 1994; Miller and Cruise, 1995; Mashriqui and Cruise, 1997). Data used during these studies were obtained from several different sources (both in situ and remote sensing) and in varying formats (digital and analog). Remotely sensed data were used in conjunction with soils and topograpic data to provide the basis for the simulation of runoff and sediment yield from the area.

Mayaguez Bay encompasses an area of approximately 100 km^2 on the west coast of Puerto Rico (Figure 16.2). As shown on the figure, three relatively small watersheds drain into the bay; the Anasco (360 km^2), the Guanajibo (311 km^2), and the Yaguez (17.4 km^2). Land-use activities within these basins include various agricultural enterprises (sugar, coffee, and dairy) as well as commercial and industrial developments. Sugar harvesting is confined to the floodplains along the main stems and major tributaries while coffee plantations are located in the upland areas. In addition, large portions of the Anasco and Guanajibo basins consist of undisturbed forested areas. The present study focused on the Guanajibo basin which is composed of agricultural and forested areas with only a small amount of urbanization.

Three streamflow gauging stations are maintained by the U.S. Geological Survey within the Guanajibo basin. Two stations are located on the main stem and one is on Rio Rosario which is the major tributary to the main channel. The Rio Rosario gauge is a continuous stage recorder and water quality station and encompasses a drainage area of 47.4 km^2. This subbasin was the initial focus of the study because in addition to discrete water quality samples, continuous suspended sediment records were available for this gauge. The simulation model was first constructed and verified for this small basin (Cruise and Miller, 1993; Mashriqui and Cruise, 1997) and then extended to the main Guanajibo watershed. Initial Guanajibo simulations covered the period 1988–1990 (Cruise and Miller, 1994), while subsequent simulations were run for the period

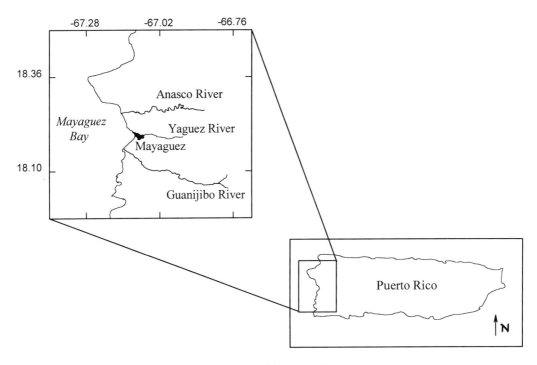

Figure 16.2. Location map: Western Puerto Rico and Mayaguez Bay.

1973–1982 (Miller and Cruise, 1995). Different sources of remotely sensed data were used to derive land cover information for the basin corresponding to these periods of time.

The northern portion of the Guanajibo watershed is composed of steep mountainous ridges (including the Rosario) which comprise about 58% of the total area. Terrain slopes within these regions vary from 32% to 56%. The less significant mountains on the southern rim (average slope = 25%) comprise about 14% of the area. The remaining 28% is comprised of the very mild floodplain area where sugar harvesting is the primary activity. Soil associations in the mountainous areas are generally clays or silty clays while silty clay loams dominate the floodplain region (Gierbolini, 1975).

Digital Data Acquisition and Processing

Data requirements for the hydrologic model included climate records, topography, soil characteristics, land use activities, and ground cover. Regional topography and hydrography (streams) were obtained from DLG files of 1:20,000 scale quadrangle maps available from the U.S. Geological Survey. Contour intervals were 10 m in areas of high relief with intermediate lines of 5 m in the coastal floodplain regions. Soil associations were hand-digitized from published soil surveys available from NRCS.

Estimates of land use and land cover were obtained for the various simulation periods using three sources of remotely sensed data: aerial photography (1975), Calibrated Airborne Multispectral Scanner data (CAMS, 1990), and a Landsat TM image (1985). The photography consisted of black and white images at a scale of 1:30,000 (U.S. Geological Survey). The photos were acquired using an antivignetting filter to minimize tonal variation within a frame. Individual photos were digitized to 256 gray scale levels using a Microtek (PC-based) desktop scanner at a 100 DPI resolution to yield a pixel ground resolution of 7.5 m. Only the portion of each photo with a consistent illumination was scanned. Twenty-six photos were required to provide complete coverage of the Guanajibo watershed. The digitized images were digitally mosaicked using the ELAS (Beverly and Penton, 1989) and Figment (Miller, 1993) image processing software. The mosaicked image was then resampled to a resolution of 30 m to correspond to the CAMS and TM data. Statistical analyses were performed on pixels within areas of known ground cover in order to determine threshold reflectance values (gray scale levels) for separation of classes. The 90th percentile point in each distribution was selected as the threshold reflectance value for that particular class (i.e., the reflectance for which 90% of the pixels were less than that value). Next, a masking routine was applied to the mosaicked image using the threshold gray scale values to separate the image into four relevant classes (e.g., highly reflectant, clear or bare ground, agriculture, and forest) for hydrologic analysis. The resulting image was georeferenced using Universal Transverse Mercator (UTM) map coordinates for the location of gauging stations (highway bridges) identified on the image. The digitized boundaries of the study areas were overlain onto the image to extract statistics and data required for the hydologic model.

Verification of the classification procedure was based on the original aerial photographs. Sample sites of approximately 50 x 50 pixels were selected at 20 random locations within the study area. These sites were then cross-referenced with the photographs. Results indicated that in 95% of the selected sites at least 80% of the pixels were correctly classified. In 50% of the cases, at least 90% of the pixels were correctly classified. Based on these results the errors in the classification procedure applied to the scanned photographs were deemed to be within acceptable limits. The classified image is shown in Figure 16.3a (see color section).

An unsupervised classification was performed on the 1985 Landsat TM data using the ERDAS software package (Figure 16.3b, see color section). A maximum likelihood classification was performed using TM channels 2, 3, and 4 with a maximum of 10 classes specified. Classes representative of hydrologically similar land cover were combined to yield class types consistent with the classification of the digitized photography. Because the TM scene had a number of clouds over the study area, these clouds and their shadows were left in separate classes and were reconciled later.

A land-cover / land-use classification of the CAMS data was created using ELAS (Figure 16.4, see color section). The CAMS data were preprocessed to remove the effects of the atmosphere using Figment. A single flight line of CAMS 30 m data was georeferenced to a UTM projection for classification. ELAS module SRCH provided unsupervised training statistics for 14 classes based on CAMS channels 2 (0.52 to 0.60 mm), 4 (0.63 to 0.69 mm), and 6 (0.76 to 0.90 mm). Using these statistics, the ELAS module CLMAXL produced an image based on a maximum likelihood classifier. The final classified image was then cross-referenced with the CAMS photography to assign specific land-use / land-cover definitions to each class. All digital data were converted to an ELAS image format for final analysis and data extraction for the hydrologic model.

Hydrologic and Sediment Yield Modeling

The hydrologic and sediment yield modeling for the Rosario and Guanajibo watersheds was accomplished using the "grouped response units" technique developed for macroscale modeling by Tao and Kouwen (1989), Kite and Kouwen (1992), and Kite et al., (1994). This method is a semi-distributed technique that makes maximum use of remotely sensed data and GIS capabilities (Cruise and Miller, 1993, 1994; Mashriqui and Cruise, 1997). Database environments utilized in this study included ELAS, ARC/INFO, and Map II (Panzer et al., 1992). The spatial database was used to represent the study area by identification of preliminary computational zones for modeling purposes based upon relative homogeneity of soils and topographic characteristics. Important soils characteristics include depths of layers corresponding to particular matrix definitions, porosity, and hydraulic conductivity. Topographic parameters include mean slope, stream density, and flow plain geometry. These characteristics were derived from the database and used to define preliminary computational zones (CZ) within the study area. Frequency histograms of hydrologically relevant properties were used to examine spatial homogeneity for CZ definition (Mashriqui and Cruise, 1997). The land cover classes were then overlain onto the computational zones and final hydrologic response units (HRU) were defined as areas of identical land cover. HRU are areas of homogeneous hydrologic characteristics such that the response to climatic forcing would be uniform (Vieux, 1988; See et al., 1992). The assumption is made that areas of equal land cover within a computational zone can be aggregated into HRU regardless of their location within the zone (Kite and Kouwen, 1992; Kite et al., 1994; Mashriqui and Cruise, 1997). Figure 16.5 (see color section) shows the computational zones for the Rosario subbasin overlain onto the land classes derived from the 30 m 1990 CAMS image. The 1985 TM image and the 1975 aerial photo image were employed in a similar manner in the Guanajibo simulations (Miller and Cruise, 1995).

A water budget model was executed once for each land class present in each computational zone (Cruise and Miller, 1993, 1994; Mashriqui and Cruise, 1997). The results from all the HRU were then summed to determine the total response from the zone. The responses from all the zones were summed in turn on a weekly basis to estimate the total basin response at the outfall point. The U.S. Agricultural Research Service model CREAMS (*C*hemical, *R*unoff, *E*rosion from *A*gricultural *M*anagement *S*ystems, Knisel, 1980) was employed for water balance accounting and sediment yield modeling using the methodology outlined above. A particular point of interest is that the CREAMS model uses the Soil Conservation Service curve number method (USDA, 1986) to separate surface runoff from infiltrated water. The curve number is significantly related to the land cover of a particular area. Research has shown that remotely sensed data, particularly TM images, can be employed with great reliability to estimate runoff curve numbers (Ragan and Jackson, 1980; Slack and Welch, 1980; Rango et al., 1983) as was done in this study. Observed runoff and sediment loading from the Rio Rosario gauge was used to calibrate and verify the initial model while runoff data were used to verify the extended Guanajibo model.

RESULTS

The results of the runoff and sediment yield simulations for the initial Rosario simulations are shown in Figures 16.6a, b and 16.6c, d, respectively (Cruise and Miller, 1993). The model was initially calibrated for the period 1986–1989 and then verified compared to observations for 1990–1991. The results of the two sets of runoff simulations for the entire Guanajibo watershed are shown in Figures 16.7 (1988–1990) and 16.8 (1973–1982), respectively (Cruise and Miller, 1994). These results demonstrate that the modeling strategy was fairly consistent in simulating runoff and sediment yield from these watersheds. As with any real-world modeling effort, some inconsistencies in model veracity are inevitable. These irregularities were due primarily to the sparse rain gauge network in the region which caused several storms to be underestimated or missed altogether in the modeling. However, taking these irregularities into account, the model appeared to respond properly to changing soil moisture regimes, thus indicating that the water budget mechanism was adequately calibrated.

The results appear to verify the modeling strategy which was largely based upon the land cover characteristics of the spatial database. The classified spectral images revealed the pattern of land use and land cover throughout the basin and aided in the subdivision of the Rosario and Guanajibo watersheds into computational zones and response units for the model. The results appear to ver-

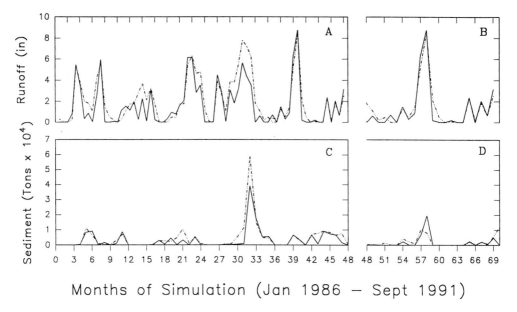

Figure 16.6A, B. Runoff simulation results versus observed data for Rio Rosario subbasin; 1986–1991; A) Calibration runs; B) Verification runs (Cruise and Miller, 1993).
Figure 16C, D. Sediment yield simulation results versus observed data for Rio Rosario subbasin; 1986–1991; C) Calibration runs; D) Verification runs (Cruise and Miller, 1993).

ify that hydrologic variables such as SCS curve numbers and vegetation characteristics can be reliably estimated from remotely sensed images. More significantly, the results support the contention that remotely sensed data from different sources can be integrated into a consistent data base for hydrologic modeling. The land-cover data were obtained from different sources corresponding to the periods of time when particular sources were available. Of particular note are the results of the 1973–1982 simulations (Figure 16.8) that were based upon the land cover obtained from the classified aerial photographs and the 1985 TM image. The simulations for this period were as accurate (r = 0.89) as were the more recent simulations (r = 0.87) for the Guanajibo based on the CAMS images (Figure 16.7).

CONCLUSIONS

Remotely sensed data integrated into a GIS environment has been shown to be a valuable asset in hydrologic modeling. In fact, recent developments have shown that remotely sensed data and GIS tools are virtually indispensable resources for modeling over large spatial domains. The development of macroscale hydrologic models, either to serve as land surface components of atmospheric circulation models, or to function in conjunction with such models, is largely predicated upon remote sensing/GIS principles. Remotely sensed data makes it possible to update hydrologic parameters at relevant temporal scales (monthly, seasonally, annually) over very large spatial domains. The increased speed, efficiency, and cost-effectiveness of GIS and image processing hardware and software have brought these capabilities down to the desktop level. In addition, the products derived from remotely sensed data, as well as the raw images themselves (many of which are available on-line at no charge) are available to virtually every hydrologic modeler. A large variety of remote sensing instruments are currently available with even more planned for launch in

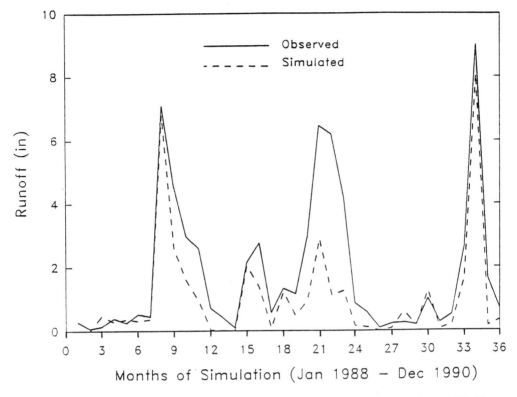

Figure 16.7. Runoff simulation results versus observed data for Rio Guanajibo, Puerto Rico; 1988–90 (Cruise and Miller, 1994).

the near future. These new hyperspectral instruments will greatly increase the use of remotely sensed data in hydrology by providing the basis for more accurate estimation of such products as soil moisture, snow cover, precipitation distribution, land cover, and vegetation biomass. Advances in hydrologic modeling methods to make use of these data at varied spatial and temporal resolutions will continue to be a major focus of research in the hydrologic sciences.

ACKNOWLEDGMENTS

A large portion of the research on hydrologic processes in Puerto Rico was conducted while the senior author was supported by a joint NASA/National Research Council Associateship at Stennis Space Center (SSC), MS. Dr. Armond Joyce, Education Affairs officer at SSC, provided assistance during this period. Additional financial support for this work was provided by NASA under grant NAGW–849 from the office of University Affairs, NASA Headquarters. Mr. Hassan Mashriqui of Louisiana State University, Baton Rouge, LA aided in classification of the Landsat TM data. We thank Jim Anderson, Greg Booth, Greg Carter, and Marco Giardino for their comments on an earlier version of this chapter.

REFERENCES

Atkinson, P.M., and A.R.L. Tatnal, 1997. Introduction: Neural networks in remote sensing. *International Journal of Remote Sensing*, 18(4): 699–710.

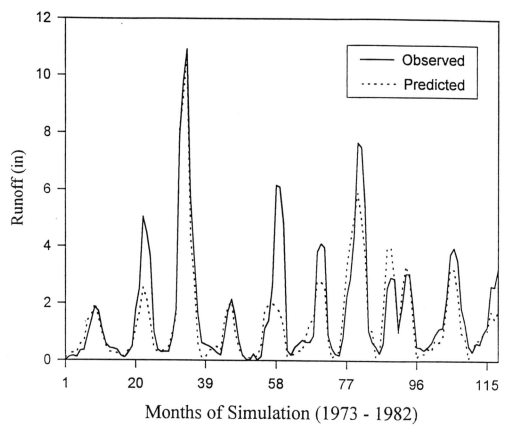

Figure 16.8. Runoff simulation results versus observed data for Rio Guanajibo, Puerto Rico; 1973–82 (Miller and Cruise, 1995).

Beverly, A.M., and P.G. Penton, 1989. ELAS Users Reference, Volume II, Unpublished Report, Science and Technology Laboratory, John C. Stennis Space Center, MS.

Bober, M.L., D. Wood, and R.A. McBride. 1996. Use of digital image analysis and GIS to assess regional soil compaction risk. *Photogrammetric Engineering and Remote Sensing*, 62(12): 1397–1404.

Boisvert, J.B., Q.H.J. Gwyn, A. Chanzy, D.J. Mayor, B. Brisco, and R.J. Brown. 1997. Effects of soil moisture gradients on modelling radar backscattering from bare fields. *International Journal of Remote Sensing*, 18(1): 153–170.

Carter, G.A., W.G. Cibula, and R. L. Miller. 1996. Narrow-band reflectance imagery compared with thermal imagery for early detection of plant stress. *Journal of Plant Physiology*, 148: 515–522.

Carter, G.A., and R.L. Miller. 1994. Early detection of plant stress by digital imaging within narrow stress-sensitive wavebands. *Remote Sensing Environment*, 50: 295–302.

Carter, G.A., 1993. Responses of leaf spectral reflectance to plant stress. *American Journal of Botany*, 80(3): 239–243.

Cialella, A.T., R. Dubayah, W. Lawrence, and E. Levine, 1997. Predicting soil drainage class using remotely sensed and digital elevation data. *Photogrammetric Engineering and Remote Sensing*, 63(2):171–178.

Courault, D., B. Aloui, J.P. Lagouarde, P. Clastre, H. Nicolas, and C. Walter, 1996. Airborne thermal data for evaluating the spatial distribution of actual evapotranspiration over a watershed in oceanic climatic conditions—application of semi-empirical models. *International Journal of Remote Sensing*, 17(12): 2281–2302.

Cruise, J.F., and R.L., Miller. 1993. Hydrologic modeling with remotely sensed databases. *Water Resources Bulletin*, 29(6): 997–1002.

Cruise, J.F., and R.L. Miller, 1994. Hydrologic modeling of land processes in Puerto Rico using remotely sensed data. *Water Resources Bulletin*, 30(3): 419–428.

Engman, E.T., 1990. Progress in microwave remote sensing of soil moisture. *Canadian Journal of Remote Sensing*, 16: 6–14.

Gierbolini, R.E., 1975. Soil Survey of Mayaguez Area of Western Puerto Rico. United States Department of Agriculture, Conservation Research Report No. 26.

Goetz, S.J., 1997. Multi-sensor analysis of NDVI, surface temperature and biophysical variables at a mixed grassland site. *International Journal of Remote Sensing*, 18(1): 57–70.

Goodrich, D.G., D.A. Woolhiser, and T.O. Keeper, 1991. Kinematic routing using finite elements on a triangular irregular network. *Water Resources Research*, 27(6): 995–1003.

Graham, R., and R. Read, 1986. *Manual of Aerial Photography*. Focal Press, Boston, MA.

Greene, R.G, and J.F. Cruise, 1996. Development of a geographic information system for urban watershed analysis. *Photogrammetric Engineering and Remote Sensing*, 62(7): 863–870.

Homer, C.G., R.D. Ramsey, T.C. Edwards, Jr., and A. Falconer, 1997. Landscape cover-type modeling using a multi-scene thematic mapper mosaic. *Photogrammetric Engineering and Remote Sensing*, 63(1): 59–68.

Jackson, T.J., J. Schmugge, and E.T. Engman. 1996. Remote sensing applications to hydrology: soil moisture. *Hydrological Sciences Journal*, 41(4): 637–647.

Jensen, J.R., 1986. *Introductory Digital Image Processing: A Remote Sensing Perspective*. Prentice-Hall, Englewood Cliffs, NJ.

Kite, G.W., and N. Kouwen, 1992. Watershed modeling using land classifications. *Water Resources Research*, 28(12): 3193–3200.

Kite, G.W., A. Dalton, and K. Dion, 1994. Simulation of streamflow in a macroscale watershed using general circulation model data. *Water Resources Research*, 30(5): 1547–1559.

Knisel, W.G., Ed., 1980. CREAMS, a Field Scale Model for Chemical, Runoff and Erosion from Agricultural Management Systems. Conservation Research Report No. 26, United States Department of Agriculture, Washington, D.C.

Lillesand, T.M., and R.W. Kiefer, 1993. *Remote Sensing and Image Interpretation*. 3rd ed., John Wiley and Sons, New York, N.Y.

Lo, C.P., D.A. Quattrochi, and J.C. Luvall. 1997. Application of high-resolution thermal infrared remote sensing and GIS to assess the urban heat island effect. *International Journal of Remote Sensing*, 18(2): 287–304.

Luvall, J.C., and H. R. Holbo, 1991. Thermal remote sensing methods in landscape ecology. In *Quanitiative Methods in Landscape Ecology*, M.G. Turner and R.H. Gardner, Eds., Springer, Verlag, New York, NY, pp. 127–152.

Maidment, D.R., D. Djokic, and K.G. Lawrence, 1989. Hydrologic modelling on a triangulated irregular network. EOS, 70(43):1091.

Mashriqui, H.S., and J.F. Cruise, 1997. Sediment yield modeling by grouped response units. *Journal of Water Resources Planning and Management*, ASCE, 123(2): 95–104.

Miller, R.L., J. DeCampo, T. Leming, and K. Hughes, 1995. Affordable solutions for analyzing coastal processes using remotely sensed data, *Proceedings of the Third Thematic Conference, Remote Sensing for Marine and Coastal Environments*, Seattle, WA, pp. 308–314.

Miller, R.L., J.C. Cruise, E. Otero, and J.M. Lopez. 1994. Monitoring suspended particulate matter in Puerto Rico: Field measurements and remote sensing. *Water Resources Bulletin,* 30(2): 271–282.

Miller, R.L., and J.D. DeCampo, 1994. C COAST: A PC-based program for the analysis of coastal processes using NOAA CoastWatch data. *Photogrammetric Engineering and Remote Sensing,* 60(2):155–160.

Miller, R.L. 1993. High resolution image processing on low cost microcomputers. *International Journal of Remote Sensing,* 14:655–667.

Miller, R.L., and J.F. Cruise, 1995. Effects of suspended sediments on coral growth: evidence from remote sensing and hydrologic modeling. *Remote Sensing of Environment,* 53: 177–187.

Murai, H., and S. Omatu, 1997. Remote sensing image analysis using a neural network and knowledge-based processing. *International Journal of Remote Sensing,* 18:655–667.

Nijssen, B, D.P. Lettenmaier, X. Liang, S.W. Wetzel, and E.F. Wood, 1997. Streamflow simulation for continental-scale river basins. *Water Resources Research,* 33(4): 711–724.

O'Neill, P., N. Chauhan, and T. Jackson. 1996. Use of active and passive microwave remote sensing for soils moisture estimation through corn. *International Journal of Remote Sensing,* 17: 1851–1865.

Otero, E., R.L. Mille, and J.M. Lopez. 1992. Remote sensing of chlorophyll and sediments in coastal waters of Puerto Rico. *Proceedings, First Thematic Conference on Remote Sensing for Marine and Coastal Environments,* New Orleans, LA.

Paine, D.P., 1981. *Aerial Photography and Image Interpretation of Resource Management.* John Wiley and Sons, New York, NY.

Panzer, M., K.C. Kirby, and N. Thies, 1992. *Map II: Reference Manual.* John Wiley and Sons, New York, NY.

Paola, J.D., and R.A. Schowengerdt, 1997. The effect of neural-network structure on a multispectral land-use / land-cover classification. *Photogrammetric Engineering and Remote Sensing,* 63(5):535–544.

Poncet, F.V., C. Prietzsch, and M. Tapkenhinrichs, 1994. Regionalization of Soil Physical Parameters Using ERS–1 PRI SAR Data. IGARSS'94.

Ragan, R.M., and T.J. Jackson, 1980. Runoff synthesis using Landsat and SCS model. *Journal of the Hydraulics Division, ASCE,* 106(HY5):667–678.

Rango, A., A. Feldman, T.S. George, III, and R.M. Ragan, 1983. Effective use of Landsat data in hydrologic models. *Water Resources Bulletin,* 19:165–174.

Rasmussen, M.S., 1997. Operational yield forcast using AVHRR NVDI data: Reduction of environmental inter-annual variability. *International Journal of Remote Sensing,* 18(5):1051–1058.

Ricotta, C., G.C. Avena, and F. Ferri, 1997. Analysis of human impact on a forested landscape of central Italy with a simplified NDVI texture descriptor. *International Journal of Remote Sensing,* 17(14): 2869–2874.

Roger, R.E., 1996. Principal components transform with simple, automatic noise adjustment. *International Journal of Remote Sensing,* 17:2719–2728.

Sabburg, J.M., 1994. Evaluation of an Australian ERS–1 SAR Scene Pertaining to Soil Moisture Measurement. IGARSS'94.

Schultz, H.,1996. Shape reconstruction from multiple images of the ocean surface. *Photogrammetric Engineering and Remote Sensing,* 62(1):93–101.

See, R.B., D.L. Naftz, and C.L. Qualls, 1992. GIS-assisted regression analysis to identify sources of selenium in streams. *Water Resources Bulletin,* 28(2):315–330.

Silfer, A.T., G.J. Kinn, and J.M. Hassett, 1987. A geographic information system utilizing the triangulated irregular network as a basis for hydrologic modeling. *Auto Carto 8, Proceedings of the 8th International Symposium on Computer Assisted Cartography*, N.R. Chrisman, Ed., American Society of Photogrammetry and Remote Sensing/American Congress of Surveying and Mapping, pp. 129–136.

Singh, V.P., and M. Fiorentino, 1996. Hydrologic modeling with GIS. *Geographical Information Systems in Hydrology*. V.P. Singh and M. Fiorentino, Eds., Kluwer Academic Publishers, Dordrecht, the Netherlands, pp. 1–13.

Slack, R.B., and R. Welch, 1980. Soil Conservation Service runoff curve number estimates from Landsat data. *Water Resources Bulletin*, 16(5): 887–893.

Smith, C., N. Pyden, and P. Cole, 1994. *ERDAS Field Guide*, 3rd ed., ERDAS, Inc., Atlanta, GA.

Sole, A., and A. Valanzano, 1996. Digital terrain modelling. *Geographical Information Systems in Hydrology*. V.P. Singh and M. Fiorentino, Eds., Kluwer Academic Publishers, Dordrecht, the Netherlands, pp. 175–194.

Su, Z., P.A. Troch, and F.E. de Troch, 1997. Remote sensing of bare surface soil moisture using EMAC/ESAR data. *International Journal of Remote Sensing*, 18: 2105–2104.

Tan, C.H., and S.F. Shih, 1988. A geographic information system for study of agricultural land use changes in St. Lucie county, Florida. *Soil Crop Science Society of Florida, Proceedings*, 47:102–105.

Tao, T., and N. Kouwen, 1989. Remote sensing and fully distributed modeling for flood forecasting. *Journal of Water Resources Planning and Management, ASCE*, 115(6): 809–823.

United States Department of Agriculture, 1986. Urban Hydrology for Small Watersheds, Technical Release 55. United States Department of Agriculture, Soil Conservation Service, Washington, D.C.

Van Deventer, A.P., A.D. Ward, P.H. Gowda, and J.G. Lyon, 1997. Using Thematic Mapper data to identify contrasting soil plains and tillage practices. *Photogrammetric Engineering and Remote Sensing*, 63(1):87–92.

Venkateswarlu, N. B., and R.P. Singh, 1995. A fast maximum likelihood classifier. *International Journal of Remote Sensing*, 16:313–320.

Vieux, B.E., 1988. Finite Element Analysis of Hydrologic Response Areas Using Geographic Information Systems. Ph.D. Dissertation, Michigan State University, East Lansing, MI.

Wegmuller, U, 1997. Soil moisture monitoring with ERS SAR interferometry. *Proceedings ERS Symposium, Florence'97*.

Wilkening III, H.A., 1989. Landsat data processing and GIS for regional water resources management in northeast Florida. *Proceedings, GIS/LIS 1989, American Congress of Surveying and Mapping*, pp. 110–119.

Wolock, D.M., and C.V. Price, 1994. Effects of digital elevation model map scale and data resolution on a topography-based watershed model. *Water Resources Research*, 30(11): 3041–3052.

Wood, E.F., M. Sivapalan, K.J. Beven, and L. Band, 1988. Effects of spatial variability and scale with implications to hydrologic modeling. *Journal of Hydrology*, 102: 29–47.

Zhang, W., and D.R. Montgomery, 1994. Digital elevation model grid size, landscape representation, and hydrologic simulations. *Water Resources Research*, 30(4):1019–1028.

CHAPTER 17

Technological Advances in Automated Land Surface Parameterization from Digital Elevation Models

Jurgen Garbrecht, Lawrence W. Martz, and Patrick J. Starks

INTRODUCTION

Topography plays an important role in the distribution and flux of water and energy within natural land surfaces. Classical examples include surface runoff, evaporation, infiltration, and heat exchange which are hydrologic processes that take place at the ground-atmosphere interface. The quantitative assessment of the processes depend on the topographic configuration of the land surface, which is one of several controlling boundary conditions. Many topographic parameters can be computed directly from a Digital Elevation Model (DEM) (Band, 1986; Jenson and Domingue, 1988; Mark, 1988; Martz and Garbrecht, 1992, 1993; Moore et al., 1991; Tarboton et al., 1991; Wolock and McCabe, 1995). This automated extraction of topographic parameters from DEMs is recognized as a viable alternative to traditional surveys and manual evaluation of topographic maps, particularly as the quality and coverages of DEM data increase. Manual evaluation of topography is general tedious, time-consuming, error-prone, and often subjective (Richards, 1981).

This chapter presents four advances in computerized methods to extract topographic parameters from DEMs. The first two advances address the treatment of depressions and flat surfaces in the DEM. Most existing methods for handling depressions and flat areas in DEM processing for drainage analysis are based on some common and fundamental assumptions about the nature of these features. These assumptions are largely implicit to the methods and are usually not recognized explicitly. These are: (1) that closed depressions and flat areas are spurious features that arise from data errors and limitations of DEM resolution; (2) that flow directions across flat areas are determined solely by adjacent cells of lower elevation; and (3) that closed depressions are caused exclusively by the underestimation of DEM elevations. While the first of these seems reasonable, the others are not. It is possible to make more reasonable assumptions about the controls on flow direction and the cause of closed depressions and to incorporate these assumptions into new algorithms for handling difficult topographic situations encountered in raster DEM processing for drainage analysis. The two new algorithms presented here are based on a deductive but qualitative assessment of the most probable nature of depressions and flat areas in raster DEM.

The last two advances address the identification of the topology of the channel network from raster images, and the parameterization of irregular overland or hillslope areas. The topology of the channel network is captured in terms of network node indexing and channel ordering by the Strahler method (Strahler, 1957). Channel ordering and node indexing is fundamental to the automation of flow routing management in distributed surface hydrology models and morphometric evaluation of channel network structure. The node index numbers can also serve to link network nodes and channel data stored in tabular format. The parameterization of subcatchments quantifies the length, width, and slope of rectangular planes representing irregularly shapes overland and hillslope contributing area. This rectangular subcatchment conceptualization is often used in distributed modeling of hillslope runoff and erosion processes. The presented subcatchment parameterization algorithms are an important contribution to traditional watershed modeling because the algorithms automate a task that is subjective and requires experience in interpretation and conceptualization of irregular hillslope features.

In the following section the fundamental principles underlying the algorithms and the essential components of the algorithms are presented. Related discussions in the broader context of water resources can be found in Garbrect and Martz (1999a). Issues relating to technical details and the implementation of these algorithms into digital land surface processing exceed the framework of this chapter. The presented technological advances are incorporated in the topographic parameterization model TOPAZ (TOpographic PArameteriZation) which automatically segments and parameterizes watersheds from DEMs for water resources, hydraulic, and hydrologic applications (Garbrecht and Martz, 1999b, 2000).

TREATMENT OF SPURIOUS DEPRESSIONS IN DEMS

Depressions are groups of raster cells completely surrounded by other cells of a higher elevation. They are usually artifacts that arise from data inaccuracies, interpolation procedures, and limited horizontal and vertical resolution of the DEM (Mark, 1983, 1988; Tribe, 1992; Zhang and Montgomery, 1994). They represent a major difficulty for DEM processing procedures that are based on the downslope flow routing concept (Martz and Garbrecht, 1992) because the existence of a downslope flow path at every cell is assumed. In the case of a depression there is, by definition, no outflow, and procedures based on an assumed downslope flow path are bound to fail.

The traditional solution to this problem is to remove all depressions in the DEM by raising the elevations within the depression to the elevation of its lowest outlet. This procedure is called "filling" of the depression. Two assumptions are implicit to this approach: (1) depressions are spurious features that arise from interpolation errors or insufficient precision in elevation values; and, (2) all depressions are due to the underestimation of elevation and should be filled. The practice of eliminating depressions solely by filling is likely to introduce systematic bias into the modified DEM. In reality, elevation errors in the DEM are as likely to result from elevation overestimation as from underestimation, and some depressions arise from the obstruction of flow paths by over-

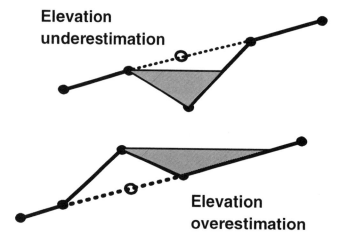

Figure 17.1. Two dimensional schematic profiles illustrating depressions arising from elevation underestimation and elevation overestimation (figure from Martz and Garbrecht, 1997).

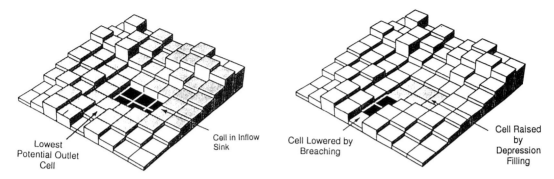

Lowest Potential Outlet Cell

Cell in Inflow Sink

Cell Lowered by Breaching

Cell Raised by Depression Filling

Figure 17.2. Breaching and filling of a spurious depression in a DEM: (2a) depression and outlet identification; (2b) outlet breaching and final filling of depression (figure from Garbrecht, Starks, and Martz, 1996).

estimated elevations (Figure 17.1). In such cases, breaching the obstruction is more appropriate than filling the depression created behind the obstruction. This breaching approach reduces or eliminates the filling of depressions created by narrow obstructions. It is particularly effective in DEMs of low relief landscapes in which obstruction of flow paths are more prevalent.

A three-step algorithm is used for breaching narrow obstructions along flow paths. First, the spatial extent of each depression and its contributing area is delineated. Second, potential outlets or overflow points on the edge of the depression area are defined. Potential outlets are those raster cells within the depression area which are: (1) adjacent to a cell outside the depression area, and (2) at a higher elevation than a cell outside the depression area. The lowest of these potential outlets is selected as the depression outlet. Third, the selected outlet is evaluated for possible lowering to simulated breaching (Figure 17.2a). The number of cells at the outlet that may be lowered by breaching is termed the breaching length. To restrict breaching to relatively narrow obstructions, the breaching length is arbitrarily set to one or a maximum of two cells. If the breaching length is longer than two cells the flow obstruction is likely to be a true topographic feature, and outlet breaching is not permitted. In the presence of a one- or two-cell breaching length, the elevation of the outlet cell(s) are lowered to the lesser elevation of the cells outside or inside of the depression at the breaching site (note: both cells outside and inside the depression at the outlet are of lower elevation per definition of the breaching length). This third step changes the elevation of the outlet and effectively breaches the obstruction responsible for the depression (Figure 17.2b). If more than one potential breaching site exists, the one with the greatest breaching depth (primary criterion), and the shortest breaching length (secondary criterion) is selected. Finally, regardless of whether a breach is performed or not, the elevations of the remaining cells inside the depression and at a lower elevation than the outlet are changed to the elevation of the outlet (Figure 17.2b). This produces a continuous flat surface at the location of the depression. A more detailed coverage of depression breaching/filling can be found in Martz and Garbrecht (1999).

TREATMENT OF FLAT SURFACES IN DEMS

Truly flat land surfaces seldom occur in nature. However in DEMs, areas of limited relief can translate into perfectly flat surfaces. Perfectly flat surfaces in DEMs can be attributed to the following three causes: (1) too low a vertical and/or horizontal DEM resolution to represent the landscape, particularly affecting low relief landscapes; (2) filling of depressions; and (3) landscape that is truly flat, which seldomly occurs. Whatever their origin, flat surfaces are problematic because flow direction on a perfectly flat surface is indeterminate (Speight, 1974; Tribe, 1992). This problem arises in automated drainage analysis with both the widely used D–8 flow routing approach (Fairchild and Leymarie, 1991) and for various multiple-direction and aspect-driven approaches (Costa-Cabral and Burges, 1994).

Traditionally, flow direction over flat surfaces in DEMs is defined using a variety of methods ranging from landscape smoothing to arbitrary flow direction assignment. For a short review of existing methods the reader is referred to Tribe (1992). Flow direction assignment over flat surfaces is particularly difficult within the framework of the D–8 method (Fairchild and Leymarie, 1991) because landscape properties are defined by the DEM cell at the point of interest and its immediate surrounding eight adjacent cells. Since all DEM cells on a flat surface have the same elevation value, a unique flow direction cannot be assigned. In the following, a generic numerical scheme is presented that allows for the identification of flow direction over flat surfaces.

This numerical scheme is based on the recognition that natural landscapes generally drain toward lower terrain while simultaneously draining away from higher terrain. This effect is incorporated into DEMs by incrementing elevations on flat surfaces to produce two gradients: one forces flow away from higher terrain; the second draws flow toward lower terrain. The selected elevation increment is arbitrarily small (say, 1 mm). Such small elevation increments are sufficient to identify flow direction over the flat surface, yet from a practical point of view they do not significantly alter the elevation of the digital land surface.

The gradient toward lower terrain is imposed by incrementing the elevation of all cells in the flat surface that are not adjacent to a cell with a lower elevation (outlet) or an existing downslope gradient. This incrementation is applied successively and repeatedly to all cells that after each incrementation pass still remain with no downslope gradient. In this way, a flow gradient toward lower terrain is constructed as a backward growth from the outlet(s) into the flat surface while at the same time satisfying all boundary conditions imposed by the higher and lower terrain surrounding the flat surface.

The gradient away from higher terrain is imposed by first incrementing the elevation of all cells in the flat surface that are adjacent to higher terrain and have no adjacent cell at a lower elevation. The imposed increment introduces a downslope gradient away from higher terrain for all cells immediately adjacent to higher terrain. In subsequent passes the incrementation is applied to all cells that have been incremented in previous passes, and also those cells that are in the flat surface and adjacent to an incremented cell, but not adjacent to a cell of lower elevation. The result of this incrementation is a gradient away from higher terrain which is grown from the edges of the higher terrain into the flat surface.

In a final step, the cumulative gradients applied in the previous two steps are linearly added for each cell to determine the total incrementation. Adding the total incrementation to the initial elevation of each cell results in a surface that is no longer flat and includes a gradient away from higher terrain, and a gradient toward lower terrain. The net effect of the elevation incrementation is the modification of elevations on the flat surface which will produce, by means of subsequent DEM processing, a flow direction pattern that is consistent with the topography surrounding the flat surface and that displays flow convergence properties. A more detailed description of drainage identification over flat surfaces can be found in Garbrecht and Martz (1997a).

The effect of the algorithm is illustrated using a saddle topography between two mountains (Figure 17.3). The saddle consists of a flat surface between higher terrain to the right and left (hatched) and three locations of lower terrain (circles). Additional complications are introduced by a wedge of higher terrain protruding into the flat surface from the bottom, and a rectangular indentation of the flat surface into higher terrain at the top right corner of the figure. The arrows show the computed drainage over the flat surface. The arrow size is proportional to the upstream

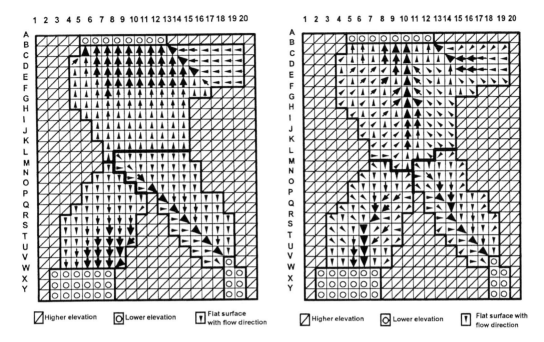

Figure 17.3. Drainage pattern over a saddle topography: (a) traditional approach; (b) new approach (figure from Garbrecht and Martz, 1997a).

drainage area of the flat surface. Figure 17.3a shows the drainage pattern produced by the model of Martz and De Jong (1988) and Jenson and Domingue (1988). This drainage pattern suffers from the "parallel flow problem" and a lack of flow convergence (Fairchild and Leymarie, 1991). Figure 17.3b shows the corresponding drainage pattern produced by the presented procedure. The drainage displays flow convergence properties and is much more consistent with the topography of the overall saddle configuration.

A second example illustrates a curved valley with a flat floor flanked by higher terrain (Figure 17.4). In addition, a small hill in the valley center creates an obstruction to drainage. The arrows show the path of the main drainage line around the inside corner of the valley bend. Drainage from behind and below the small hill converge rapidly and join the main drainage line. Any tributary from the higher valley sides would enter the flat surface, follow the indicated arrows, and join the main drainage line within a short distance. The flow convergence and drainage pattern in Figure 17.4 is reasonable, given that the initial valley floor was flat and contained no topographic information to guide the drainage identification.

NETWORK AND SUBCATCHMENT INDEXING

Once the channel network and direct contributing areas are automatically defined from DEMs, they are usually displayed as a raster image which consists of strings and groups of raster cells with numeric codes or colors that distinguish the network and subcatchments. For these images to be useful in watershed management and runoff modeling, individual channel links and contributing areas must be explicitly identified and associated with topological information for upstream and downstream connections. Such identification is often possible in vector GIS, but usually not in raster GIS. An algorithm that can analyze images of raster channel networks, index network nodes,

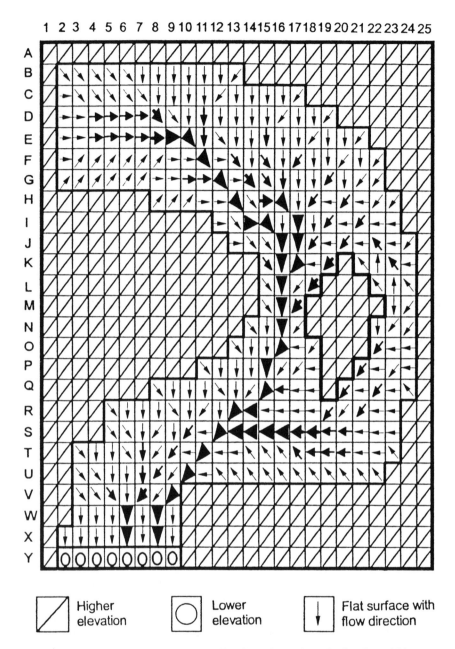

Figure 17.4. Drainage pattern on a flat valley floor (figure from Garbrecht and Martz, 1997a).

and order the channels by the Strahler method is presented. This algorithm provides a direct link between GIS images and hydrologic models, and leads to automated processing of segmented watersheds by distributed hydrologic models.

It is assumed that an image of an unidirectional, fully-connected network has already been defined. Four steps are required to fully identify the topology of the network and subcatchments. In the first step, flow direction information at each raster cell is used to move cell by cell along the channels of the network from upstream to downstream to determine the Strahler order (Strahler,

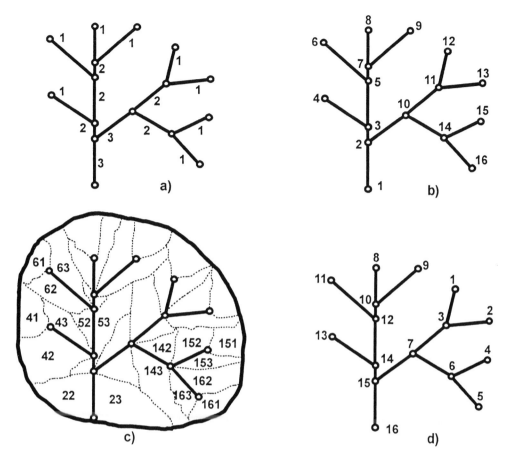

Figure 17.5. Drainage network and node indexing: (a) Strahler orders; (b) node indexing; (c) sub-catchment indexing for selected node numbers 2, 4, 5, 6, 14, 15 and 16; (d) sequence in which the channels are to be processed for flow routing (figure from Garbrecht and Martz, 1997b).

1957) of each channel link and to identify the location of all source and junction nodes (Figure 17.5a). Flow direction is determined as the steepest downslope flow path from the current cell to one of the eight neighboring cells. In the second step, the node location and flow direction information are used to simulate a walk along the left bank of the channel network beginning and ending at the watershed outlet (Croley, 1980). The walk begins at the watershed outlet which is assigned node number 1. Then the walk traces the channel that ends at the outlet to the next node upstream. At that node the left channel is followed further upstream. As each node is passed during this walk, it is assigned the next sequential node number (Figure 17.5b). When a source node is encountered the walk moves in the downstream direction until a partially evaluated node is encountered, at which time the node-by-node walk is resumed again in an upstream direction along the unevaluated channel branch.

In the third step, the subcatchment areas for each channel link are identified. These consist of direct contributing areas into the left-bank, right-bank and, in the case of exterior links, into the source of the channel link. Subcatchments are assigned an identification code based on the previously assigned node numbers. For all subcatchments, a base identification number is assigned which is the node number (NN) at the upstream end of the link to which the subcatchment drains multiplied by 10 (NN*10). For source node subcatchments a value of 1 is added to (NN*10), for left-bank subcatchments a value of 2 is added, and for right-bank subcatchments a value of 3 is added. Channel cells (which are not considered to be part of the right-bank, left-bank, or source node subcatchments) are assigned an identification code of (NN*10) plus 4 (Figure 17.5c).

These identification codes provide a basis on which network links, nodes, and subcatchments can be associated with one another. Most importantly, in a fourth step the node numbering scheme can be used to determine the optimal routing sequence to be used in modeling streamflow through large and complex networks (Garbrecht, 1988) (Figure 17.5d). The algorithm makes possible the automated quantification of network structure from raster network images and greatly enhances the direct linkage of GIS-generated channel networks and hydrologic and hydraulic models. Further details on this algorithm can be found in Garbrecht and Martz (1997b).

OVERLAND AREA PARAMETERIZATION

Automated identification of landscape parameters for individual overland areas within a subdivided watershed is the next step in automated DEM processing for hydrologic/hydraulic model application (Goodrich and Woolhiser, 1991). Overland areas are defined as undissected hillslopes of irregular shape that drain directly into a channel link. For hydrologic modeling these overland areas are often approximated by a rectangular plane of given width (W), length (L) and slope (S) (Wooding, 1965; Smith at al., 1995; Feldman, 1995). Such a Wooding representation of a subcatchment consists of two rectangular planes joined to form a V-shaped valley along which the stream flows (Figure 17.6). The geometric dimensions W, L, and S of the plane are essential for the determination of the magnitude, shape, and timing of the overland runoff hydrograph, and are used in models such as KINEROS (Smith et al., 1995) and the kinematic option of HEC–1 (Feldman, 1995). Three numerical expressions are presented that identify the plane parameters to reproduce the hydraulic runoff characteristics of the original overland areas. These expressions are implemented using flow path, accumulated area and elevation data derived from the DEM raster. This raster data can be obtained by suitable DEM processing software such as the digital landscape analysis tool TOPAZ (Garbrecht and Martz, 1999b, 2000).

The definition and assumption necessary for the first algorithm are: (1) a flow path is the route traveled by the water from an upstream source to the channel at the downstream edge of the overland area; and, (2) flow paths with large discharge contribute proportionally more to the runoff hydrograph characteristics than flow paths with small discharge. Based on this definition and assumption the model for hydraulically representative plane length (L) can be formulated as a discharge-weighted mean length of all flow paths within the irregular overland area. Furthermore, the discharge on hillslopes is often proportional to upstream area, and the expression for plane length can be formulated as follows:

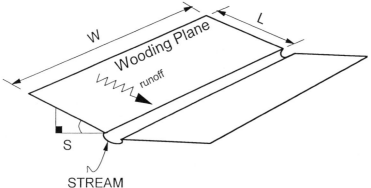

Figure 17.6. Wooding catchment representation (figure from Garbrecht, Martz, and Goodrich, 1996).

$$L = \frac{\sum_{i=1}^{n} l_i * a_i}{\sum_{i=1}^{n} a_i} \qquad (1)$$

where l is flow path length, a is upstream drainage area, and subscript i is flow path counter.

The second numerical expression defines the plane width and is derived from area and water conservation considerations. To ensure water continuity and area conservation, plane width is computed as the overland area (A) divided by the plane length (L) computed by Eq. 1.

Finally, the expression for plane slope is a weighted average of all flow path slopes. Flow path slope is defined as the change in elevation from the top to the bottom of the flow path divided by flow path length. All flow paths in the irregular overland areas are considered. The expression for the plane slope is given by:

$$S = \frac{\sum_{i=1}^{n} s_i * k_i}{\sum_{i=1}^{n} k_i} \qquad (2)$$

where s is flow path slope and k is a weighting factor. The weighting is defined in one of three ways: (1) upstream drainage area weighted with $k_i = a_i$; (2) flow path length weighted with $k_i = l_i$; or, (3) upstream drainage area times flow path length weighted with $k_i = a_i * l_i$.

In the first case, length weighting is used because slope estimates of long flow paths are generally more accurate since the elevation and length values have been obtained over a larger number of discrete raster units, effectively reducing the raster resolution noise. Length weighting also favors flow paths with larger drainage areas (i.e., larger discharge) because drainage area and flow path length are often related. In the second case, drainage area weighting is used to emphasize flow paths with larger drainage areas which contribute proportionally more to the runoff hydrograph characteristics than those with smaller drainage areas. The third case is a combination of the first and second case with the product of length and drainage area as the weighting factor. Until further research establishes the most appropriate method for the estimation of plane slope, the results of each of the three methods should be considered in the determination of the final plane slope.

A more detailed discussion of the overland area parameterization can be found in Garbrecht et al. (1999).

CONCLUSIONS

Advances in the treatment of depressions and flat surfaces, the identification of network topology, and the parameterization of overland areas are presented. The proposed depression removal by a combination of breaching and filling, as well as the gradient imposition of flat surfaces, produce a more realistic and consistent drainage pattern than traditional depression filling and local drainage searches over flat areas. The presented improvements are particularly important for drainage and erosion investigations.

The identification of channel network and subcatchment topology from raster images provides an important linkage between DEM-derived drainage features and automated watershed management and hydrologic modeling. Finally, the length and slope of the rectangular overland area approximation are computed as drainage area weighted averages over the length and slope of all flow paths in the irregular overland areas. The strength of the procedure lies in its consideration of the geometric properties of each contributing flow path rather than relying on a lumped approach, and in the consideration of qualitative cause-effect relations between flow path and runoff characteristics to emphasize flow paths that are more important to the runoff process.

REFERENCES

Band, L. E., 1986. Topographic partition of watersheds with digital elevation models. *Water Resources Research,* 22(1):15–24.

Costa-Cabral, M. C., and S. J. Burges, 1994. Digital Elevation Model Network (DEMON): A model of flow over hillslopes for computation of contributing and dispersal areas. *Water Resources Research,* 30(6):1681–1692.

Croley, T. E., 1980. A micro-hydrology computation ordering algorithm. *Journal of Hydrology,* 48:211–236.

Fairchild, J., and P. Leymarie, 1991. Drainage networks from grid digital elevation models. *Water Resources Research,* 27(4):29–61.

Feldman, A. D., 1995. HEC–1 flood hydrograph package, Chapter 4 of *Computer Models of Watershed Hydrology,* Singh, V. J., Ed., Water Resources Publication, Highlands Ranch, CO, pp. 119–150.

Garbrecht, J., 1988. Determination of the execution sequence of channel flow for cascade routing in a drainage network. *Hydrosoft,* 1(3):129–138.

Garbrecht, J., L. W. Martz, and D. C. Goodrich, 1996. Subcatchment parameterization for runoff modeling using digital elevation models. In *Proceedings of the American Society of Civil Engineers Hydraulics Conference,* North American Water and Environment Congress '96, ASCE, Anaheim, CA, 24–28 June, 1996.

Garbrecht, J., P. J. Starks, and L. W. Martz, 1996. New digital landscape parameterization methodologies. In *Proceedings of the 32nd Annual Conference and Symposium on GIS and Water Resources, American Water Resources Association,* Herndon, VA, TPS–96–3, September 22–26, 1996, Fort Lauderdale, FL, pp. 357–365

Garbrecht J., and L. W. Martz, 1997a. The assignment of drainage direction over flat surfaces in raster digital elevation models. *Journal of Hydrology,* 193:204–213.

Garbrecht J., and L. W. Martz, 1997b. Automated channel ordering and node indexing for raster channel networks. *Computers and Geosciences.* 23:961–966.

Garbrecht J., and L. W. Martz, 1999a. TOPAZ: An Automated Digital Landscape Analysis Tool for Topographic Evaluation, Drainage Identification, Watershed Segmentation and Subcatchment Parameterization, Overview, ARS-GRL 99–1, U.S. Department of Agriculture, Agricultural Research Service, Grazinglands Research Laboratory, El Reno, OK, 26 pp.

Garbrecht, J., and L.W. Martz, 1999b. Digital elevation model issues in water resources modeling. In *Hydrologic and Hydraulic Modeling Support in GIS,* D. Maidment and D. Djokic, Eds., ESRI Press, Redlands, CA, 216 pp.

Garbrecht, J. and L. W. Martz, 2000. TOPAZ: An Automated Digital Landscape Analysis Tool for Topographic Evaluation, Drainage Identification, Watershed Segmentation and Subcatchment Parameterization, TOPAZ User Manual, ARS-GRL, 2-00, U.S. Department of Agriculture, Agricultural Research Service, Grazingland Research Laboratory, El Reno, OK.

Garbrecht, J., D.C. Goodrich, and L. W. Martz, 1999. Method to quantify distributed subcatchement properties from digital elevation models. In *Proceedings of the 19th Annual AGU Hydrology Days,* Atherton, CA, pp. 149–160.

Goodrich, D. C., and D. A. Woolhiser, 1991. Catchment hydrology. *Reviews of Geophysics*, Supplement, April 1991, pp. 202–209, U.S. National Report to International Union of Geodesy and Geophysics. 1987–1990.

Jenson, S. K., and J. O. Domingue, 1988. Extracting topographic structure from digital elevation data for geographical information system analysis. *Photogrammetric Engineering and Remote Sensing,* 54(11):1593–1600.

Mark, D. M. 1984. Automatic detection of drainage networks from digital elevation models. *Cartographica*, 21 (2/3):168–178.

Mark, D. M., 1988. Network models in geomorphology. In Anderson, M. G., Ed., *Modeling Geomorphological Systems*. John Wiley and Sons, Chichester, UK pp. 73–96.

Martz, L. W., and E. De Jong, 1988. CATCH: A FORTRAN program for measuring catchment area from digital elevation models. *Computers and Geosciences*, 14(5):627–640.

Martz, L.W., and J. Garbrecht, 1992. Numerical definition of drainage networks and subcatchment areas from digital elevation models. *Computers and Geosciences* 18(6):747–761.

Martz, L. W., and J. Garbrecht, 1993. Automated extraction of drainage network and watershed data from digital elevation models. *Water Resources Bulletin*, 29(6):901–908.

Martz, L. W., and J. Garbrecht, 1998. The treatment of flat areas and depressions in automated drainage analysis of raster digital elevation models. *Hydrologic Processes* 12:843–855.

Martz, L. W., and J. Garbrecht, 1999. An outlet breaching algorithm for the treatment of closed depressions in a raster DEM. *Computers and Geosciences,* 25:835–844.

Moore, I. D., R. B. Grayson, and A. R. Ladson, 1991. Digital terrain modeling: A review of hydrological, geomorphological and biological applications. *Hydrological Processes*, 5(1):3–30.

Richards, N. S. 1981. General problems in morphology. In Gordie, A. S., Ed., *Geomorphological Techniques*. Allen and Umnin, London, pp. 26–30.

Smith, R. E., D. C. Goodrich, D. A. Woolhiser, and C. L. Unkrich, 1995. KINEROS—A kinematic runoff and erosion model. Chapter 20 of *Computer Models of Watershed Hydrology,* V. J. Singh, Ed., Water Resources Publication, Highlands Ranch, CO, pp. 697–732.

Speight, J. G., 1974. A Parameteric Approach to Landform Regions. Special Publication of the Institute of British Geographers, 7:213:230.

Strahler, A.N., 1957. Quantitative analysis of watershed geomorphology. *Trans. AGU*, 38(6): 913–920.

Tarboton, D. G., R. L. Bras, and I. Rodriguez-Iturbe, 1991. On the extraction of channel networks from digital elevation data. *Water Resources Research*, 5:81–100.

Tribe, A., 1992. Automated recognition of valley lines and drainage networks from grid digital elevation models: A review and a new method. *Journal of Hydrology*, 139:263–293.

Wolock, D. M., and G. J. McCabe, 1995. Comparison of single and multiple flow direction algorithms for computing topographic parameters in TOPMODEL. *Water Resources Research*, 31(5):1315–1324.

Wooding, R. A., 1965. A hydraulic model for the catchment-stream problem. *Journal of Hydrology*, 3:254–267.

Zhang, W., and D. Montgomery. 1994. Digital elevation model grid size, landscape representation and hydrologic simulations. *Water Resources Research*, 30(4):1019–1028.

CHAPTER 18

Aerial Photointerpretation of Hazardous Waste Sites: An Overview

Donald Garofalo

INTRODUCTION

Much interest has been expressed recently by scientists interested in the capabilities and limitations of various remote sensors for characterizing hazardous waste and waste disposal sites. Special attention has been focused on the ability of these sensors to locate and characterize waste and waste site activities at abandoned facilities. As such, sophisticated new radar systems, thermal sensors, and other multispectral and hyperspectral scanners are being assessed as potential candidates for addressing the waste discovery and characterization problem.

Airborne radars can penetrate sand in hyperarid environments, between 1–6 meters in depth (Sabins, 1987); penetration is minimal in humid soil areas. Thermal sensors can display temperature variations at the surface, sometimes suggestive of subsurface conditions and multispectral and hyperspectral scanners can detect subtle changes in the surface environment which may be suggestive of buried features. The historical aerial photograph, however, frequently has resolutions for detecting barrel size or smaller features, and the ability to show site conditions long ago. Thus, the historical aerial photograph, unlike many newer and more sophisticated sensors, is an inexpensive source of invaluable information which can be used to locate a waste disposal site and measure accurately the size and dimensions of currently buried or overgrown features and to generally track the history of waste disposal site activity from beginning to end.

THE HISTORICAL AERIAL PHOTOGRAPH

The historical aerial photograph is the sensor of choice of the Environmental Protection Agency's Environmental Photographic Interpretation Center (EPIC), which for the past 25 plus years has been applying this tool for locating potential waste disposal sites and characterizing these sites and associated waste disposal practices. The historical aerial photograph is an extremely powerful remote sensing tool. It is the only remotely sensed data to have recorded events at sites frequently as far back in time as the 1930s. This is of immense value to the Superfund program which is charged with looking for and evaluating abandoned hazardous waste disposal sites, sites which today may display no evidence at the surface of their former use (Slonecker et al., 1999). The tool is of substantial value in litigation as evidence of the past waste disposal practices of PRPs (Principal Responsible Parties), and has been highly successful at assisting EPA and Department of Justice lawyers in winning their cases. In addition, the ability to view the aerial photo stereoscopically (in 3-D) and to measure, using photogrammetry, the heights, depths, volumes, and other dimensions of features and materials currently present, long removed from, or buried at

a site has also contributed invaluable evidence in courtroom situations, and has helped to reconstruct pictorially for jurors and judges alike and in an easily understandable format, the activity at
a site over time. Finally, the availability of historical aerial photographs from both government and
private sources, and the extent of aerial photo coverage of the United States, through various federal agency mapping programs such the Agricultural Stabilization and Conservation Service (now
Farm Services Agency), and the U.S. Geological Survey, to name only two, have literally ensured
that a hazardous waste disposal site located within the conterminous US has been overflown and
site conditions documented more than once during the site's disposal history.

This chapter focuses on the value of the historical aerial photograph and the various image
analysis and mapping functions which use aerial photos for analyzing hazardous waste disposal
sites.

THE BASICS OF AERIAL PHOTOINTERPRETATION FOR
WASTE SITE CHARACTERIZATION

Backlighting and Variable Magnification

Using backlighted tables with adjustable illumination, and high-power magnification zoom
stereoscopes, aerial photographs are interpreted for site size, drainage patterns, type of fill materials, leachate, burial sites, lagoons, impoundments and their contents, and general condition of the
site. Locations and descriptions of tanks, drums, open storage areas, evidence of vegetation stress,
on-site obstacles, structures, equipment, access routes, and other details may also be obtained
through photo analysis. Historical analysis provides the information necessary to obtain a chrono-
logical understanding of a site's development and activities. This information is particularly important for describing and illustrating past activities and conditions at abandoned hazardous waste
disposal sites which fall under the jurisdiction of the Comprehensive Environmental Response,
Compensation, and Liability Act (aka Superfund) program.

Film Transparencies

The use of aerial photo transparencies on backlighted variable illumination tables maximizes
the available information content of analyzed aerial photographs. Aerial photo transparencies are
first generation copies of originally exposed film. Each additional step of film processing, such as
producing additional film copies or photographic prints degrades from the original product and reduces the amount of information contained in the original photo. High-powered magnifying
scopes are used to identify subtle, but often significant features on aerial photos which can easily
be overlooked if not viewed with the benefit of backlighting and variable magnification.

Stereoscopy

The importance of stereoscopy in the photointerpretation process cannot be ignored. Stereoscopy allows the photo analyst to see features on an aerial photograph in three dimensions (3-D).
Through stereoscopic parallax (the apparent displacement of the position of a feature in an image
caused by a change in the position of observation) a stereoscope may be used to view overlapping
aerial photographs to provide a three-dimensional effect of features on the ground (Figure 18.1).
When coupled with various measuring devices such as stereo comparators or other digital photogrammetric devices highly accurate measurements can be made of the dimensions (height,
widths, lengths, depth) of features seen on the aerial photograph. Volumes of materials and volu-

Acquisition of Stereoscopic Aerial Photographs and Stereoscopic Parallax *

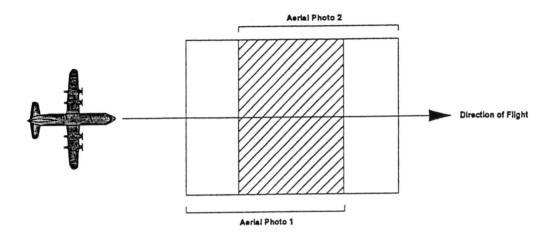

Area of 60% Forward Lap
Area of Forward Lap represents the same ground area photographed along a flightline by two overlapping aerial photos.

* Stereoscopic parallax is the displacement of the position of a feature in an image caused by a shift in the observation position. A stereo pair are two overlapping photos containing stereoscopic parallax, and which may be viewed using a stereoscope to provide a three-dimensional effect. The same effect can be obtained with a hand-held camera by taking a picture of the same object from two separate locations and then viewing the pair of pictures together with a stereoscope (Sabins, 1987).

Using special photogrammetric equipment accurate measurements can be made on stereo photo pairs. Contour maps which show differences in terrain elevation are also made using this technique.

Dimensions of features such as their length, width, and height, as well as volumetric measurements such as amounts of mounded materials, or depths and sizes of excavations can be calculated using photogrammetric tecniques.

Figure 18.1. Acquisition of stereoscopic aerial photographs and stereoscopic parallax. The area of 60% forward overlap is shown as the cross-hatched area, and it represents the same ground area photographed along a flightline by two overlapping aerial photos.

metric capacities of excavations can be calculated. This same technique is used to make topographic maps which show land surface elevations using contour lines.

The Signature Concept

A basic part of photointerpretation is the extraction of useful information from an image. Although this may be performed by human (manually) or machine (electronically by computer) depending on the form of the original data (e.g., as analog, hard copy photos versus digital images, respectively) the photo analysis performed by EPIC in support of EPA's hazardous waste program is conducted almost exclusively by manual, not machine analysis. Using manual methods, some types of information displayed on aerial photos are obvious to anyone used to reading a map. An example includes large bodies of water. However, the vast bulk of information is not evident to the untrained or inexperienced viewer.

The training and experience that makes feature identification possible on aerial photos is based on learning to recognize combinations of imagery characteristics called *signatures*. A *signature* is a combination of visible characteristics (such as color, tone, shadow, texture, size, shape, pattern, and association) which permit a specific object or condition to be recognized on an aerial photograph. The relative potential of a given signature to enable possible, probable, or positive identification of an object or condition is the *degree of certainty*. *Possible* is a term signifying a degree of certainty of signature identification when only a few characteristics are discernible on a photo or these characteristics are not unique to a signature. *Probable*, on the other hand, is a term signifying a degree of certainty of signature identification when most characteristics, or strong or unique characteristics of a signature are discernible but fall short of positive identification.

When interpreting an aerial photograph, the analyst is generally searching for the signature of one or more objects or conditions by viewing aerial photo stereopairs through stereoscopes. An analyst relies either on experience or "ground truth" information (corroborative information obtained through other data sources or on-site field visits to the site) in identifying signatures. When working with historical aerial photographs of areas that have changed or are inaccessible, experience becomes critically important. For some forms of photo-interpretation, e.g., to map wetlands vegetation or forest types, signatures representative of various vegetation types are confirmed by on-site visits and then extrapolated over a much larger area to map vegetation types displaying the same signatures, but not visited on the ground.

The *signature* concept is not an all or none concept since signatures can vary in degree of certainty. Because of this it is essential to clearly distinguish between positive identifications and calls of lesser certainty. Because hazardous waste site characterization by aerial photo interpretation may be used to support civil or criminal litigation, it is important that the degrees of certainty be clearly stated by the photo interpreter.

The characteristic signature of a given object or condition can vary with the type of film or imagery, scale, resolution, and other factors. Therefore, aerial photos vary in suitability depending on the object of the analysis. For example, 55-gallon drums can be positively identified on average quality, 1:6,000 scale aerial photographs. On the other hand, typical 1:20,000 scale aerial photos do not normally allow for positive identification of drums, but may allow for *possible* or *probable* identifications to be made. Even smaller scale (higher altitude) imagery is so inappropriate for drum identification that even *possible* identifications cannot be made. Variability in photo type also affects the interpretability and signatures of objects. Natural color, color infrared, and black and white aerial photos, for example, have no sensitivity for heat detection, while thermal scanner imagery can reveal temperature differences of objects. An experienced photo interpreter is able to use a variety of remote sensing tools and is fully knowledgeable of the capabilities *and* limitations

Table 18.1. Hazardous Waste Disposal Site Features/Activities Routinely Extracted by Image Analysis from Historical Aerial Photographs

Access road	Extraction area	Pressure tank
Berm	Feature boundary	Probably underground drainage
Building	Feature outline	Railroad
Channelized drainage	Fence	Refuse
Chemical storage	Fenced site boundary	Revegetated
Cleared area	Fill	Revetment
Container	Flow direction	Site boundary
Crates/boxes	Graded area	Sludge
Culvert/bridge	Ground scar	Solid waste
Cylindrical object	Historical boundary	Stacked objects
Dark-toned	Horizontal tank	Stain
Debris	Impoundment	Standing liquid
Dike	Indeterminate drainage	Structure
Disposal area	Lagoon	Study area
Disturbed ground	Landfill	Surface runoff
Drainage	Leachate	Suspected drainage
natural	Light-toned	Tank farm
channelized	Liquid	Tank trailer
suspected or historical	Material	Trench
indeterminate flow	Medium-toned	Unfenced site boundary
natural	Mounded material (extensive)	Vegetated
channelized	Mounded material (small)	Vegetation stress
tidally influenced	Objects	Vehicle
natural	Open storage	Vehicle access
channelized	Outfall	Vertical tank
Drums	Pipeline	Waste disposal area
Edge of slope	Pit	Wastewater treatment plant
Excavation	Pond	Wetland
Excavation, pit (extensive)	Possible drum area	

of these tools for specific applications. Table 18.1 is a listing of the kinds of features, natural resources, and site activities which are routinely identified on aerial photographs by skilled image analysts for characterizing waste disposal sites.

OVERVIEW

Hazardous Waste Site Analyses

Hazardous waste disposal site characterization using historical and current aerial photographs comprises a major part of EPIC's workload. Utilizing the vast archives of aerial photographs of the country maintained by government and private sources, dating back to the 1930s, EPIC's analysts reconstruct the waste handling and disposal history of a site in order to support site cleanup and regulatory or enforcement efforts. Aerial photographs have proved to be powerful tools in court in the form of evidence and to support expert witness testimony and for facilitating the recovery of millions of dollars in site cleanup costs and penalties from responsible parties (Garofalo and Wobber, 1974; Erb et al., 1981; Evans and Mata, 1984; Stohr et al., 1987; Mata and Christie, 1991).

The information interpreted from an aerial photograph is annotated onto a clear film overlay which identifies and delineates the location of significant ground features and activities. Accom-

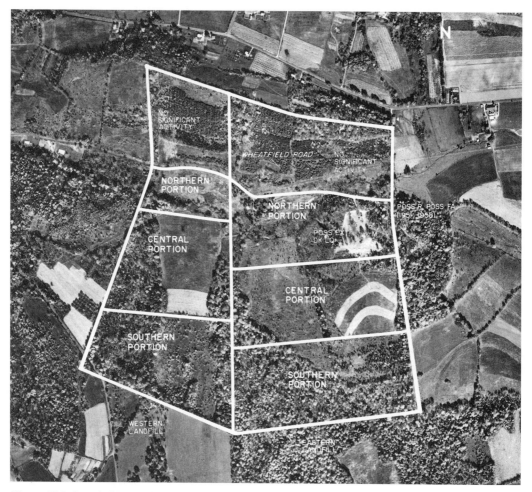

Figure 18.2. Sample historical site analysis prepared by EPIC. October 19, 1958. No significant activity was evident on the 1946 or 1947 photographs (not shown here). Significant features observed on the 1951 photographs are annotated and discussed in conjunction with the following analysis.

The landfill study area consists of two landfills, the Western Landfill and the Eastern Landfill. Both landfills are south of Wheatfield Road. No significant activity is visible within the portion of the study area north of Wheatfield Road throughout the analysis, and this area will not be discussed further.

Each landfill is divided into three portions for the purpose of discussion. The northern, central, and southern portions will not be annotated further. Fill areas are annotated in both landfills throughout the analysis, but are not discussed individually. Disposal activities noted within fill areas are annotated and discussed.

Drainage at and around the landfill is shown in the Wetland and Drainage analysis (Figure 18.10). Minimal changes were evident in the overall drainage routes throughout the years of analysis. Minor variations in drainage resulted from the gradual development of the landfill. These transient drainage routes are annotated for each year of analysis, but are not discussed unless a significant change is visible.

Western Landfill. No significant activity is evident.

Eastern Landfill. Northern portion. In 1951, possible refuse (R) was noted in a possible fill area (FA) outside the eastern site boundary.

In 1958 the east side has been partially cleared (not annotated). Material (M) is piled in three areas within the clearing. A possible excavation (EX) with dark-tone (DK) liquid (LQ) is noted on the south edge of the clearing. Possible refuse remains evident in the possible fill area outside the eastern site boundary.

Central and southern portions. No significant activity is evident.

Figure 18.3. June 17, 1964. Western Landfill. No significant activity is evident.

Eastern Landfill. Extensive clearing (not annotated) is underway in the eastern landfill, and brush piles (not annotated) are scattered throughout the clearing.

Northern portion. The material and possible excavation with liquid seen in 1958 are no longer visible. Piles of coarse-textured (CT) material, an open storage (OS) area, a building (B) with a parking area, and an empty excavation, possible a catchment basin, are noted in the northern portion of the landfill. A variety of objects and equipment are visible in the open storage area. The open storage area, building and parking area remain active and expand throughout the analysis, and will be annotated but not discussed in detail.

The possible refuse seen outside the eastern site boundary in 1958 is no longer visible.

Central Portion. Piles of coarse-textured material and a rectangular bermed area are visible. The bermed area appears empty, and its use is not evident.

Southern Portion. An impoundment (IM1) and an excavation are visible. IM1, a leachate collection impoundment, is dry inside and contains a small amount of material. The excavation contains murky liquid.

Figure 18.4. April 17, 1968. Western Landfill. North portion. A wetland has been cleared (CL); however, there is no evidence of fill activity in this location.

Central Portion. Extensive filling has occurred since 1964, an impoundment (IM2), an excavation, grading (GR), and material are evident. A channel leads from the east edge of the landfill into IM2. Murky liquid is visible in the channel and in IM2. The excavation is on the southwest side and appears to be empty.

Southern portion. No significant activity is evident.

Eastern Landfill. Northern portion. The coarse-textured material noted in 1964 is no longer present. A berm and the grading of a fill area are evident in the northern portion. The graded fill area includes the former excavation which was noted in 1964.

Central portion. The bermed area and debris (DB) are visible. The bermed area remains empty, and the debris is along the edge of a fill area.

Southern portion. IM1, debris, two pits, and two excavations are visible. IM1 contains liquid, the debris is along the edge and sides of a fill area, the northern pit is full of dark-toned liquid, and the southern pit contains a small amount of dark-toned liquid/material. Both excavations contain murky liquid.

A drainage channel is visible at the south edge of the site. It receives runoff from slopes south and east of the landfill as well as from on-site. The channel leads west to the natural drainage, which flows north (Figure 18.10).

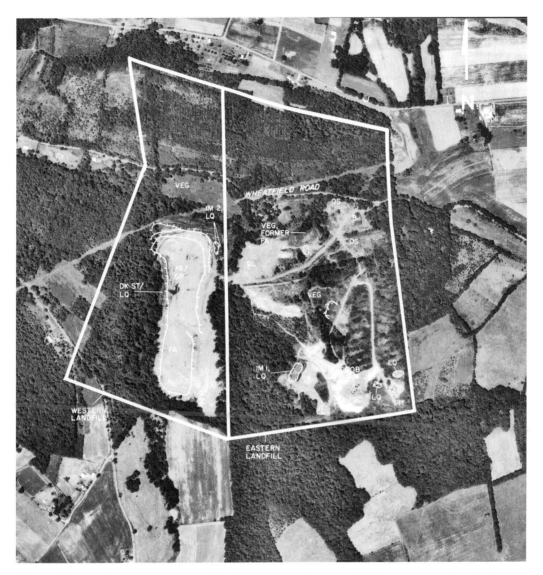

Figure 18.5. July 18, 1971. Western Landfill. North portion. Vegetation (VEG) is growing where the clearing of a wetland was noted in 1968.

Central Portion. Filling has continued since 1968, and the upper surface of the landfill has been graded. IM2 contains dark-toned liquid. Dark-toned liquid is adjacent to the east side of IM2, indicating the possibility of a breach or overflow in the impoundment. The western excavation is no longer visible. A dark-toned stain (ST) extends from the top of the landfill downslope to the base, as if caused by a liquid.

Southern portion. Extensive filling has occurred since 1968. A clearing is noted along the southeast edge.

Eastern Landfill. Northern portion. The fill area with a berm noted in 1968 is vegetation. A shallow excavation is noted south of the former fill area. The graded fill area noted in 1968 and part of the parking area around the building are in use for open storage. A large clearing is seen on the west side of the landfill.

Central Portion. Vegetation is growing where a fill area with debris was noted in 1968. The bermed area appears inactive and is partially overgrown with vegetation. A clearing is seen on the west side of the landfill.

Southern portion. The northern pit seen in 1968 is no longer present; the southern pit is smaller and appears empty. IM1 contains liquid. The debris seen along the edge of a fill area in 1968 is no longer visible, and the top of the fill area has been graded. A small amount of probable debris is evident on top of the fill area. The two excavations remain and contain murky liquid.

Tree canopy obscures the drainage channel seen along the south edge of the landfill in 1968.

panying text provides a full site description. Figures 18.2 through 18.11 illustrate a standard historical site analysis of a hazardous waste disposal site, and show the kinds of information that can
be extracted from an analysis of these photos, and changes at the site over time. The historical
analysis clearly shows and documents the location of major fill areas which were evident during
the life of the landfill. Disposal-related activities identified within the fill areas include debris, material, refuse, dark-toned liquid in erosion rills, excavations, and pits containing liquid. Between
1964 and 1984 a leachate collection impoundment was present on the site and a second leachate
collection impoundment was present from 1968 through 1984. An upgraded leachate collection
impoundment was present from 1980 through 1992.

Inventories of Potential Hazardous Waste Sites

Inventories of potential hazardous waste sites covering large areas and decades in time are a
very cost-effective way to discover sites for future investigation. The aerial photographs are systematically searched for specific features or to identify types of sites. Type might include landfills,
open dumps, scrap salvage yards, chemical handling and storage facilities, impoundments, or
abandoned industrial sites. Identified sites are located on overlays to topographic maps accompanied by data sheets describing site conditions. The site conditions are presented chronologically
with the period of site activity shown on the map overlay. This approach is helpful to determine
the origin of a progressive problem and to identify a hazardous site that is currently hidden by new
development.

Emergency Response

EPIC reacts through quick response capabilities to emergency situations such as hazardous material releases and natural disasters like hurricanes (Hugo) and earthquakes (San Francisco/Oakland,
CA). Aerial photographs are flown, processed and analyzed to provide immediate information to on-
site personnel regarding circumstances not easily or safely observed from the ground. Typical products for an emergency response include an immediate telephone report to on-site personnel followed
by photographs or positive film transparencies with interpretation results annotated on overlays, annotated topographic maps, and a short letter report describing analysis results.

Wetlands

Wetlands analyses are performed by EPIC in support of various sections of the Clean Water Act
concerning enforcement, permitting, and advance identification. Analysis of historical aerial photographs is often the only means of establishing the prior existence of wetlands on lands that have
been dredged or filled, and for calculating wetlands loss acreage necessary for mitigation settlements. Aerial photographs also provide information concerning vegetative type, periodicity of
flooding, tidal influences, and affected drainage patterns.

EPIC image analysts perform various types of wetlands mapping depending on the needs of the
requesting EPA headquarters or regional program office. A *wetlands/upland* boundary delineation
is the simplest form of mapping performed (Figure 18.10). The purpose for this delineation is to
identify the location of wetlands and to separate wetlands from nonwetlands areas. A *detailed wetlands analysis* is performed to identify and classify various wetland types, often using the wetlands and deepwater habitat classification developed for the Fish and Wildlife Service by
Cowardin et al. (1979). Using *jurisdictional wetlands delineation* procedures and associated fieldwork, EPIC maps wetlands in support of Section 404 of the Clean Water Act which protects wet-

lands from unpermitted dredge and fill activities. Also, in support of the *Advance Identification* process of Section 404, wetlands maps are prepared as a cost-effective way to identify wetlands in advance of permit application and evaluation. EPIC also performs image interpretation for wetlands delineation in support of EPA enforcement cases. Historical aerial photographs are used along with field checking to map wetlands losses and change due to filling and/or dredging activities. Photogrammetry is used to make accurate quantitative measurements of wetlands losses, extent of filling, and overall area changes. Geographic Information Systems (GIS) technology is applied to produce maps and graphics displays which are suitable for courtroom presentations and which clearly illustrate these changes.

Photogeology

EPIC employs aerial photointerpretation to study the geology of an area from an analysis of landforms, drainage, tones/textures, and vegetation distribution. EPIC conducts two types of photogeologic analysis and mapping: fracture trace and lithologic.

Fracture trace analysis involves the use of aerial photographs and other types of remote sensing imagery to identify linear features on the earth's surface that are naturally occurring and are surface manifestations of subsurface fracture zones in the bedrock (Figure 18.9A). When viewed in cross-section, fracture traces are seen to be vertical or near vertical breaks in the bedrock. Fractures are of particular environmental concern because contaminants are likely to move more easily through zones of fractured bedrock than through the surrounding more consolidated bedrock material. Thus, fracture traces can be used to identify possible migration routes of pollutants and are often used in the placement of monitoring/remedial wells around hazardous waste sites (Stohr et al., 1987; Scheinfeld et al., 1988; Mata and Christie, 1991).

EPIC conducts *lithologic mapping* (mapping of distinct rock types or units) from aerial photographs in order to produce a more accurate geologic map in areas where geologic mapping is incomplete due to limited fieldwork, small map scale, or other factors (Figure 18.9B). This procedure is usually performed by consulting available geologic maps of the area.

Photogrammetric Mapping

In support of EPA's mission, EPIC produces highly accurate topographic and planimetric maps, generally at a large scale, which conform with National Map Accuracy Standards and EPA Photogrammetric Mapping Specifications. Map scales, contour intervals, and planimetric details can be varied to suit specific requirements.

EPIC uses analytical stereoplotters, digital video plotters, or other digital photogrammetric methods to measure the area and volume of hazardous wastes; determine the height and placement of containment berms, dikes, and impoundments (Figure 18.8); and determine the depth of waste pits. Changes in size, shape, and other physical characteristics of a waste site are documented through sequential photogrammetric mapping (Slima, 1980).

Photogrammetric techniques are also used by EPIC to establish precise location and orientation data to support geophysical monitoring or for monitoring well placement.

Geographic Information Systems (GIS)

EPIC applies GIS technology to support a variety of EPA Headquarters or Regional Program Office Needs. For example, a National Priorities List (NPL) hazardous waste site investigation was performed using information from diverse sources (including numerous years of historical

Figure 18.6. April 7, 1981 Significant features seen on the 1980 photograph are annotated and discussed with the following analysis on page 231.

Figure 18.6, continued. Western Landfill. Central portion. In 1980 IM2 contained a small amount of liquid. In 1981 there is an increase in the amount of liquid in IM2. The road (not annotated) on the east side of the landfill has been widened. A shallow pit, at the bottom of which appears to be a drain, possibly associated with the leachate collection system, is visible on the northwest side of the widened road.

Southern portion. In 1980 a clearing was visible on the west side. In 1981 the clearing (not annotated) is expanded and graded. A slight increase in the elevation indicates fill activity since 1980. Debris (not annotated) is evident along the edges of the fill area. Several piles of earthen-toned material are visible on the south side of the fill area, and in the clearing along the southeast edge.

Eastern Landfill. Northern portion. In 1980 the leachate collection system was visible in the large clearing where grading was seen in 1975. The leachate system consisted of IM3, IM4 and IM5, all of which contained liquid. One large and one smaller pile of material were seen in a former fill area which was used for open storage in 1975. This may have been excess soil dug out while excavating the leachate collection impoundments. An increase in open storage was noted around the buildings.

In 1981 IM3, IM4, and IM5 contain liquid. Both piles of material are graded. There has been a further increase in open storage since 1980.

Central portion. In 1980 extensive filling was evident. Dark-toned liquid was pooled in two areas on the west side. Partial clearing occurred along the east side. In 1981 filling is in progress in a new area. A vehicle (V) and refuse are visible in the new fill area, indicating current activity. The east side has been completely cleared and grading is in progress.

Southern portion. In 1980 extensive filling had occurred since 1975 and the east edge was partially cleared. Dark-toned liquid was visible in numerous erosion rills on the west side of the fill area. IM1 contained liquid, and its edge was more well-defined than in previous years, suggesting that the impoundment was dredged and redug. An access road led to a clearing east of the study area. Disturbed ground (DG) was noted along a second access road which led southwest of the study area, turned southeast, and continued to a wooded area where debris/refuse was evident. The same access road also continued southwest, and debris/refuse was evident at the edge of a second wooded area (not shown). The two wooded areas south of the study area are also accessible from roads outside of the landfill.

In 1981 further filling has taken place. The east edge is completely cleared and grading is in progress. The dark-toned liquid seen in 1980 is no longer visible. There is a significant increase in the grade east of IM1, which contains liquid. Refuse is abundant in the southeast corner.

The access road and clearing remain east of the study area. Disturbed ground remains evident along the access road seen south of the study area, and debris/refuse remains visible in both wooded areas.

The drainage channel previously seen along the south edge of the landfill is visible and contains liquid.

Figure 18.7. April 17, 1988. Significant activity observed on the 1983, 1984, and 1987 photographs is discussed, but not annotated in conjunction with the following analysis on page 233.

Figure 18.7, continued. Western Landfill. Central portion. In 1983 IM2 contained liquid, and the pit seen northwest of the widened road in 1981 was no longer visible. In 1984 IM2 contained liquid, and a small clearing was noted at the end of an access road. In 1987 IM2 was no longer visible, trees had been cut around the area where it was previously located, and revegetation was evident. The clearing seen at the end of an access road in 1984 was no longer visible. In 1988 no significant activity is noted.

Southern portion. In 1983 the clearing seen in 1981 was vegetated except for a small portion on the south side. The debris and material were no longer visible, and the clearing seen at the southeast edge in 1981 was vegetated. In 1984 the southern clearing remained and no further significant activity was evident. In 1987, an access road led west from the southern clearing to a ground scar at the creek. Clearings were seen on the southwest and southeast edges; debris was noted in the southwest clearing. In 1988 linear objects (LO), probably uprooted trees, are piled in the southern clearing. A series of objects, possibly straw bales for erosion control (not annotated), are noted between the southern clearing and a long clearing. This is the same location where an access road and ground scar were noted in 1981. Refuse is seen in the southeast clearing, and linear objects, probably uprooted trees, and refuse are seen in the southwest clearing.

Eastern Landfill. Northern portion. In 1983 IM3, IM4, and IM5 contained dark-toned liquid. Disturbed ground/light-toned material was noted southwest of the office buildings. In 1984 IM3, IM4, and IM5 contained dark-toned liquid. IM6 had been constructed and was empty. The disturbed ground/light-toned material was no longer visible. Clearing and fill activity were evident southeast of IM3, IM4, and IM6. In 1987 IM5 was full of dark-toned liquid, IM4 was partially filled with dark-toned liquid, IM3 appeared empty, and IM6 was surrounded by vegetation and contained dark-tone liquid. Extensive filling and grading occurred in the area southeast of the impoundments. In 1988 IM5 and IM6 contain dark-toned liquid and IM3 and IM4 contain medium-toned liquid. The vegetation has been cleared from around IM6 and the edges appear higher. No further filling or grading is evident in the area southeast of the impoundments.

Central portion. No significant change was noted in 1983 or 1984. In 1987 extensive filling and grading had occurred. In 1988 a series of excavations, possibly catchment ponds, containing murky liquid and linear objects, probably uprooted trees, are noted on the west side.

Southern portion. In 1983 IM1 remained and contained liquid. The access road and clearing remained active east of the study area. In 1984 IM1 remained and contained liquid, and trees were cleared west of IM1. The access road and clearing remained active east of the study area. In 1987 IM1 was no longer present. Disturbed ground was evident in the area where it was previously located. The access road and clearing east of the study area were vegetating. In 1988 no significant change is evident. Debris/refuse remains visible in both wooded areas south of the study area. Between 1984 and 1987 the drainage channel along the south edge of the site was rerouted northward through a culvert as shown in the Wetlands and Drainage Analysis (Figure 18.10). It continues west to the natural drainage, which flows north.

Figure 18.9a. Introduction. This fracture trace analysis presents findings based on the study of aerial photographs and geologic literature (see References) for the area on and around the landfill.

Fracture traces and other geologic lineaments are considered to be the surface expressions of vertical to near vertical zones of fracture concentration in bedrock. Fracture trace analysis is the technique of using aerial imagery for locating fracture traces or geologic lineaments on the earth's surface based on photogeologic signatures such as soil-tonal variations and vegetational and topographic alignments.

The landfill and surrounding area are astride the boundary separating the Great Valley section of the Valley and Ridge Physiographic Province. The bedrock geology of the landfill and surrounding area consists of Paleozoic and Triassic age sedimentary rocks, Triassic age igneous rock, and assorted contact metamorphosed rock.

Findings. Of particular note are the upper Cambrian carbonates (Millback Formation) which straddle the northern section of the site (Figure 18.9B). The Millback carbonates, reported to be highly permeable (well yields in the 225 gpm range), are surrounded by rock (diabase, meta sandstones and shales) having significantly lesser yields. Because of this, contaminants from the site may have difficulty moving out of the Millback aquifer, possibly concentrating in the Millback.

Efforts to characterize the groundwater regime around the landfill should in part focus on delineating the areal extent of the Millback (mapped by MacLachlan et al. as inferred). Although a photogeologic analysis of presite imagery (1946, 1947, and 1951) basically corroborates the MacLachlan et al. map, a more detailed field analysis is warranted.

Groundwater flow through the bedrock in this area is predominantly through secondary openings such as frac-

tures, and in the Millback perhaps through solutionized conduits. Identification of fracture traces is important when modeling aquifer contamination because the fracture traces may act as conduits for concentrated groundwater flow. Fracture trace analysis can be used to map possible pathways along which contaminants may migrate away from sites via subsurface routes. Once these possible pathways have been mapped, a monitoring well network can be designed to take into account the fracture pattern identified during the photogeologic analysis.

Figure 18.9b, above. A total of 25 fracture traces was identified using 1946, 1947, and 1951 imagery. They are presented on a photographic enlargement of the 1980 aerial photograph (Figure 18.9a). Of note are fracture traces labeled A, B, and C. A and B bisect the landfill area and therefore have a high likelihood of carrying contaminants. Fracture trace C, oriented north-south, is located along the expected direction of regional groundwater flow away from the site (west-north-west), and could deflect the groundwater flow away from this expected route.

Secondary openings in the Millback carbonates may be solutionized, resulting in an enhanced groundwater flow. Depending on the degree of solutionization, any cones of depression developed around wellheads can greatly influence groundwater flow direction. Caution should be exercised if fracture traces are to be used in the location of monitoring wells. Differences of a few feet can determine whether or not a well is located on a fracture trace; therefore, it is important that a geology field team and a photogeology analyst with stereo aerial photographic transparencies and related interpretation equipment be on site for precise placement of the monitoring wells or well network.

Recommendations. Based on the findings of this photogeologic analysis, future groundwater characterization efforts at the landfill should be directed toward: (1) defining the areal extent of the Millback Formation, (2) placing monitoring wells on fracture traces A, B, and C, and (3) determining the degree of solutionization in the Millback Formation.

Figure 18.10. Wetlands and drainage analysis. Wetlands (W), open water (OW), uplands (U), and drainage are annotated for the area at and around the landfill using photographs from April 17, 1988. The general direction of drainage flow is north and west through a series of unnamed streams which eventually lead into the north-flowing Main Creek (not shown). The on-site drainage flows through a series of wetlands that contain emergent and forested vegetation.

aerial photographs, geological data, digital line graph data, soil data, property ownership, monitoring well data, etc.). These data were combined to produce topical maps and analyses for use in the Remedial Investigation/Feasibility Study decision-making process under Superfund. Another project in support of the Environmental Monitoring and Assessment Program (EMAP) involved the ecological characterization of a pilot site using photo-derived information on land use, vegetation,

Figure 18.8. Landfill Area Measurements from Photogrammetry of Leachate Collection Impoundments.

YEAR	IM1	IM2	IM3	IM4	IM5	IM6
1971						
Hectares	0.143	0.073	not	not	not	not
Acres	0.353	0.181	present	present	present	present
1981		no				
Hectares	0.108	significant	0.095	0.145	0.077	not
Acres	0.267	change	0.235	0.359	0.189	present
1988						
Hectares	not	not	0.116	0.166	0.097	0.172
Acres	present	present	0.287	0.409	0.240	0.424
Increase (+) or						N/A
Reduction (−)		no				(one year
Hectares	−0.035	significant	+0.021	+0.021	+0.020	measured)
Acres	−0.083	change	+0.052	+0.050	+0.051	

Area measurements were obtained for impoundments 1–6 using photographs representing the year of peak landfill activity (1971), an intermediate year (1981), and the year of final closure (1988). The impoundments were digitized from photographic prints using an XY digitizing table. The digitizing table coordinates were transformed into the UTM coordinated system to facilitate area measurements, using a GIS.

IM1 was visible from 1964 through 1984, and decreased in size by 0.035 hectare (0.083 acre) between 1971 and 1981. IM2 was visible from 1968 through 1984, and was 0.073 hectare (0.181 acre) in 1971. Changes in the size of IM2 were minimal, and therefore have not been measured. IM3, IM4, and IM5 were visible from 1980 through 1992, and were each enlarged by approximately 0.021 hectare (0.05 acre) between 1981 and 1988. IM6 was visible from 1984 through 1992, and was 0.172 hectare (0.424 acre) in 1988.

wetlands, and landforms to produce maps and overlays for landscape characterization and trend analysis (Norton and Slonecker, 1990).

Global Positioning Systems

Global positioning systems (GPS) technology is used by EPIC primarily to produce accurate latitude/longitude coordinates for sites under investigation, and as a means to evaluate and quantify the spatial accuracy of digital map data, and to create accurate cartographic products. In 1990 EPA adopted its Locational Data Policy (LDP) for the purpose of ensuring the collection of accurate, fully documented locational coordinates as part of all agency-sponsored data collection activities. An accuracy goal of 25 meters was established and GPS is considered the best method for achieving this goal. GPS is used to provide accurate latitude/longitude ground control points for producing accurate maps from aerial photographs and to precisely locate hazardous waste disposal sites (Wells et al., 1986; Slonecker and Groskinsky, 1993).

Miscellaneous Analyses

Additional image analyses performed by EPIC in support of EPA needs include interpretation of thermal infrared imagery for detection of illegal river discharges and landfill and mine fires;

Figure 18.11. Vegetation stress analysis, September 24, 1992. A color infrared overflight was acquired for the landfill during full leaf-on conditions to facilitate the vegetation stress analysis. The photo shown here is a black and white copy of the color infrared image which normally shows vegetation as various shades of red.. Vegetation stress (V) can be caused by many interrelated factors such as insect infestation, disease, drought, water inundation due to natural and/or man-made changes in hydrology, and soil compaction, as well as contaminants. Often vegetation can be weakened by one factor and consequently become susceptible to one or more of the others. Due to the abundant rainfall and below-normal temperatures throughout the eastern United States during the summer of 1992, it is unlikely that vegetation stress has been caused by drought conditions.

Since individual dead trees are likely to be naturally occurring snags, they have not been identified. In some cases dead or stressed vegetation may have been overcome with climbing vines, resulting in a healthy vegetation signature.

A standard or "normal" signature was developed for the vegetation at the landfill by studying the areas at and around the site and establishing a generalized spectral reflectance pattern. Possible vegetation stress was identified based on deviations from the "normal" vegetation signatures and is discussed for the western and eastern landfills.

and aerial photointerpretation for detection of abandoned oil, gas, and water wells; mapping of submerged aquatic vegetation; and land use and drainage mapping.

Aerial Photo Acquisition

EPIC acquires historical photographs, dating to the late 1930s, from a wide range of federal, state, and local government agencies and private aerial survey companies. EPIC sets the specifications for overflight of new aerial photography, to meet the Agency's varying needs. New aerial photography is acquired through a network of private aerial survey companies across the country. These companies are also available on short notice for emergency response efforts.

SUMMARY

EPIC is a facility of the Landscape Ecology Branch, Environmental Sciences Division, National Exposure Research Laboratory, Office of Research and Development, and has been performing aerial photointerpretation and mapping in support of EPA headquarters and program offices since the early 1970s. Over the years the technology of remote sensing and mapping has advanced dramatically. EPIC has maintained pace with this technology and today provides the agency with the benefits derived from its use. An evolution has also been occurring with regard to environmental concerns. The health and well-being of our nation's ecological resources, and concerns about global climate changes have added a new dimension to the way we view the environment, and require methods for efficiently collecting data inexpensively and over large areas. Advances in remote sensing and mapping technology provide important tools to help us learn more about these regional and global problems. EPA's EPIC is poised and prepared to take on these new challenges through the next century.

Notice: The U.S. Environmental Protection Agency, through its Office of Research and Development (ORD) funds and performs the work described here. This document has not been sub-

Western Landfill. The overall vegetation patterns appear healthy, and a few small patches of irregular vegetation growth (not annotated) are evident. 1. Possible vegetation is noted in the most recently active clearing, which is located in the southeast corner of the landfill. Irregular patches of vegetation are growing within barren areas, and vegetation along the adjacent natural drainage route (Figure 18.10) appears stressed. The healthiest vegetation within the clearing is growing on top of the remaining probable uprooted trees noted in 1988.

Eastern Landfill. In general, the vegetation is growing irregularly, with healthy patches interspersed with possible stressed patches (not annotated), and numerous areas which are void of vegetation. 2. Possible vegetation stress is noted along the drainage route at the south edge of the landfill, and continuing west to the natural drainage (Figure 18.10). The lowlands along drainage routes are of particular importance because they offer a natural pathway for the migration of contaminants. It was not possible to determine deviations in the vegetation signature of the on-site wetlands due to the small size and number of these areas (Figure 18.10). A generalized spectral reflectance pattern can only be established when a number of similar areas within the site vicinity can be compared to each other.

The most effective way to verify possible vegetation stress and to investigate causes of vegetation stress is with a combination of aerial photograph analysis and a field check of the study areas during the active growing season.

jected to the Agency's review; therefore, it does not necessarily reflect the views of the EPA. The U.S. government has the right to retain a nonexclusive, royalty-free license in and to any copyright covering this article.

REFERENCES

Cowardin, L., V. Carter, F. Golet, and E. LaRoe, 1979. Classification of Wetlands and Deepwater Habitats of the United States. U.S. Department of Interior, Fish and Wildlife Service, FWS/OBS–79/31.

Erb, T., W. Philipson, W. Tang, and T. Liang, 1981. Analysis of landfills with historical airphotos. *Photogrammetric Engineering and Remote Sensing*, 47:1363–1369.

Evans, B., and L. Mata, 1984. Aerial photographic analysis of hazardous waste disposal sites. In *Hazardous Wastes and Environmental Emergencies*. HMCRI, Houston, TX.

Garofalo, D., and F. Wobber, 1974. Solid waste and remote sensing. *Photogrammetric Engineering and Remote Sensing*, 40:45–49.

Mata, L., and G. Christie, 1991. The use of remote sensing techniques in environmental assessments at CERCLA and RCRA facilities. In *Proceedings of the Eighth Thematic Conference on Geological Remote Sensing*, Vol. 2, Denver, CO, pp.931–944.

Norton, D., and E. Slonecker, 1990. The environmental monitoring and assessment program's landscape characterization database: New opportunities in spatial analysis. In *Proceedings of GIS/LIS'90*, Anaheim, CA.

Sabins, F., 1987. *Remote Sensing Principles and Interpretation*. W. Freeman Co., New York, NY.

Scheinfeld, R., L. Mata, P. Landry, and A. Logue, 1988. *Use of Fracture Trace Analysis in Thickly Mantled Karst Terraces and Their Solutions*. National Water Well Associate, Dublin, OH, pp.183–210.

Slima, C., 1980. *Manual of Photogrammetry*. American Society of Photogrammetry, Falls Church, VA.

Slonecker, T., and B. Groskinsky, 1993. Environmental applications of GPS technology. In *Proceedings of the Federation Internationale Des Geometres*, New Orleans, LA.

Slonecker, T., M. Lacerte, and D. Garofalo, 1999. The value of historical imagery. *EOM*, 8:39–41.

Stohr, C., W. Su, P. DuMontelle, and R. Griffin, 1987. Remote sensing investigations at a hazardous waste landfill. *Photogrammetric Engineering and Remote Sensing*, 53:1555–1563.

Wells, D., N. Beck, D. Dlikaraoglou, A. Kleusberg, E. Krakiwsky, G. Lachapelle, R. Langely, M. Nakiboglu, K. Schwarz, J. Tranquilla and P. Vanicek, 1986. *Guide to GPS Positioning*. Canadian GPS Associates, Fredericton, NB, Canada.

CHAPTER 19

Remote Sensing and GIS for Site-Specific Farming

John G. Lyon, Andrew Ward, Bruce C. Atherton,
Gabriel S. Senay, and Tom Krill

INTRODUCTION

There are a number of capabilities that have been postulated to be useful in agriculture. These capabilities have been realized through numerous efforts over the years (e.g., Moran et al., 1997). The focus now is to demonstrate the capabilities of available technologies for operational and practical applications.

The hope is that research can demonstrate that available remote sensor, GPS, and GIS technologies can supply good information for management of crops, soils, and waters on a within-field basis, and do so at reasonable cost. The goals are to potentially improve yields while maintaining or improving soil tilth and water quality (Ward and Elliot, 1995; Blackmer and Schepers, 1996; Gowda et al., 1999).

Precision agriculture approaches can be implemented using a suite of technologies. To address within-field management concerns it is necessary to navigate over short distances and small elevations. The advent of Differential Global Positioning Systems (Van Sickle, 1996) now provides for the required precision and accuracy. The need for mapping the spatial data collected through a variety of means is met by position information from GPS and mapping capabilities of Geographic Information Systems (GIS)(Lyon and McCarthy, 1995). GIS supplies the maps of soil grid sampling, management applications, and on-the-go yield, and can do so for each season and each year of evaluation with high accuracy (Bolstad and Smith, 1995). To supply detail as to crop, soil, and hydrological conditions over the growing season and from year to year, remote sensor data from aircraft or spacecraft are employed (Eidenshink and Haas, 1992; Thenkabail et al., 1992, 1995; Ward and Elliot, 1995; Yang et al., 1998). Use of different parts of the electromagnetic spectrum can help to separate types of crops and weeds, general soil and soil moisture characteristics, and the potential influences of hydrology on soils and crops (Huete and Tucker, 1991; Lyon et al., 1998).

Our work over the last 12 years and the work of others helps to illustrate the experimental and operational capabilities of remote sensor, GPS, and GIS technologies for agriculture. Here, the background to these methods, their results, and options for the future are described and discussed using our experience from three experiments on commercial farms in Ohio, and the experience of other researchers.

BACKGROUND

In this chapter we focus on the variables which are most useful in site-specific farming and can be best related to remotely sensed data. These techniques and methods can be incorporated with GIS and watershed models to answer important water resource questions. In general, the following variables can be measured in fields and related to remotely sensed measures of light: wet and dry biomass, crop residue, leaf area index (LAI), plant population density, areal cover of crop canopy, concentration or quantity of chlorophyll, leaf tissue nutrient content, crop height, plant moisture content, yield after harvesting, on-the-go yield, soil moisture content, soil texture, soil fertility, soil nitrogen, soil phosphorus, soil potassium, soil water release characteristics, the presence and location of weeds, wet and droughty areas, and plant stresses associated with diseases, insects, or other factors.

Crop Characteristics

The wet, green biomass of healthy plants can be measured in a number of ways remotely using the visible, near, and middle infrared regions of the electromagnetic spectrum (Singh, 1989; Thenkabail et al., 1994; Senay et al., 2000a, 2000b). A number of people make use of the fact that green plants absorb most of the red light from sunshine for photosynthesis, and reflect very little red light to the sensor above the crop (Price, 1992). In addition, green plants absorb little near infrared light, but reflect great quantities of it to the sensor above. This differential light reflectance can be used as a tool to identify green plants from the background materials including soils, crop residue, and water.

Conversely, plants that do not perform optimally or are "stressed" will often have less chlorophyll and be chloretic or yellow (Blackmer and Schepers, 1996). This decrease in chlorophyll can be detected by a decrease in red light absorbance and infrared light reflectance. This differential light reflectance can be used as a tool to identify green plants from the background materials including soils, crop residue, and water (Figure 19.1).

Residue

An important consideration to many farm management activities is the maintenance of crop residue cover during the nongrowing season. The reasons for this practice include decreasing soil erosion from water and wind, reduction of plowing costs, decreasing water quality problems downstream, and others. The implementation of crop residue cover or conservation tillage practices has been of great interest, as have methods for measuring the extent of this practice (Messer et al., 1991; Jakubauskas et al., 1992). We have conducted work on measuring crop residue to track the implementation of conservation tillage, and to better understand residue influences on spectral responses during the growing season. This work began in the early 1980s and continues today (Thenkabail et al., 1992, 1994; Van Deventer et al. 1997; Gowda et al., 1999).

Crop residue or senescent plants reflect different amounts of light as compared to green plants or soil. It is possible to identify crop residue because of these differences, that include the absence of chlorophyll, presence of nonchlorophyll plant pigments, difference in water holding capabilities as compared to live or green plants, different leaf or stalk structure as compared to live plants, and different water holding, pigment, and structural characteristics as compared to soils. These differences manifest themselves as a distinct spectral signature or differential light reflectance that can be measured in the visible, near, and middle infrared regions of the spectrum (Ward and Elliot, 1995).

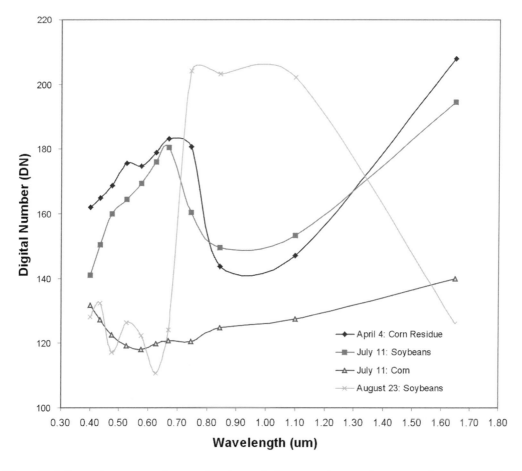

Figure 19.1. Spectral responses of corn and soybeans during vegetative growth stages, soybeans at maturity and corn residue.

Soils

The reflectance characteristics of soils are related to the parent materials, the texture, the moisture content, organic matter, and to a certain extent slope and elevation. Soils are generally lighter or brighter toned than plants and water. The exceptions would include soils of high organic matter or low bulk density or high moisture content. Bright tones of soils can be distinguished from plant materials in the visible and infrared portions of the spectrum. Work on using remote sensing to study a wide range of soil properties has been conducted by many researchers (Hatfield and Pinter, 1993; Moran et al., 1997)

An important consideration is that a dense crop canopy will obscure the bright tone of soils. This supplies a ready method to make estimates of canopy closure or leaf area index due to the darkness of the plant biomass obscuring the soil background. Conversely, measures of the presence of bare soil can help assess the absence of crop or crop canopy that has not developed (Lyon et al., 1986).

Water in a Crop Environment

Water characteristics of plants and soils can be identified and differentiated from other materials due to the fact that water or wet soils or wet plants have a lower reflectance of light to the sensor above (Lyon and McCarthy, 1995; Ward and Elliot, 1995). In general, the presence of water in any concentration decreases the light reflectance of materials. This is well known for the case of soils, where the presence of either water at the surface, water saturation, or standing water will greatly reduce the generally relatively bright reflectance of dry or relatively dry soils (Lyon, 1993).

A major use of remote sensor data is in the evaluation of crop moisture conditions. In particular, the middle infrared (approximately 1.5 and 2.2 µm) and to a certain extent the near infrared (approximately 0.7 to 1.1 µm) supply good detail as to relative plant moisture conditions of the leaves and stalk.

METHODS

Several experiments were developed over time to test the capabilities of remote sensor data for precision agriculture. The results of those experiments are presented here and in other publications. In general, the experiments included: evaluations of satellite and ground measurements of crop, soil, and hydrology variables for more than 50 commercial farms in Seneca County of north central Ohio (Thenkabail et al., 1992, 1994; Van Deventer et al., 1997); evaluations of crop, soil, hydrology, and weather characteristics using fine spatial resolution (1 m picture elements or pixels) and moderate spectral resolution (12 spectral bandwidths) for a commercial farm in Pike County of south central Ohio used in the Midwest Systems Evaluation Area (MSEA) studies (Senay et al., 1998, 2000a, and 2000b); and evaluations of low-cost aerial photographs taken from low altitudes of four commercial farms and one experimental farm station in Van Wert County of west central Ohio, and the MSEA site. Details on the methods to collect and analyze field data for these experiments are reported in several publications (Thenkabail et al., 1992, 1994; Nokes et al., 1997; Senay et al., 1998).

Landsat Thematic Mapper Data

During a three-year period, ground and satellite data were collected and analyzed for commercial farms in Seneca County (Thenkabail et al., 1992, 1994; Van Deventer et al., 1997). The goal was to evaluate the utility of Landsat satellite Thematic Mapper (TM) data for crop, soil, residue, and hydrology characteristics. This large experiment involved many field data collections during the growing season as detailed elsewhere. For remote sensing purposes, the research team visited the county on a biweekly basis to collect data during the Landsat satellite overpasses. The result was a very rich data set to evaluate field and satellite measured characteristics of crops, soils, and management practices.

Landsat 5 TM data and now Landsat 7 ETM+ data provides 30 m by 30 m resolution data every 16 days or so, and some type of Landsat data has been available for 25 years. It provides information in the visible, infrared, and thermal wavelengths.

Airborne Multispectral Scanner Data Collection

Several overflights of the MSEA site in Pike County were conducted in 1994. The platform was an Aero Commander twin engine aircraft with a camera bay holding the Daedalus Multispectral Scanner (MSS) instrument (model 1260) and an aerial camera by Wild (RC–8). The overflights were made on April 4, July 11, August 15, August 23, and September 15, and corresponded to

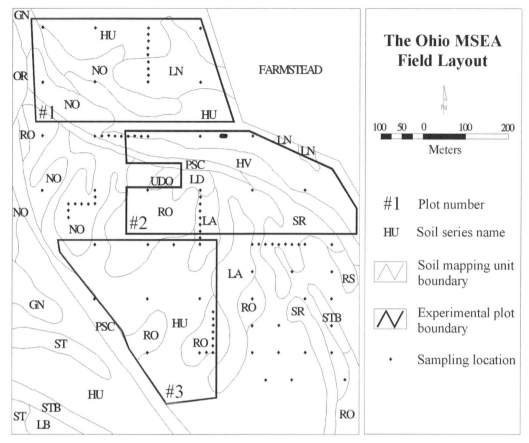

Figure 19.2. Ohio MSEA soil map (Wu et al., 1996). Source: US Department of Agriculture, Natural Resources Conservation Service, Columbus, Ohio.

early spring and preplanting, early planting and germination, mature crop, and senescent soybean, and early senescent corn crop growth periods, respectively. Ground data were collected close in time to the overflights, by a crew that measured the MSEA on a regular basis in support of a number of studies, and by personnel concerned specifically with the overflights.

Other remote sensor data were collected during 1994 and have been analyzed. These data included airborne radar data in X, C, and L-bands from a Lockheed P–3 Orion aircraft and sensor supplied by the Navy, and the hyperspectral sensor AVIRIS flown by a NASA U–2 aircraft.

The MSEA effort is ongoing, and is significant because of the extensive data collection effort in the field. In addition, products such as a Digital Elevation Model (DEM) were developed from a Differential Global Positioning System (GPS) experiment, on-the-go yield measurements were made from a combine with GPS capability, and a very extensive soil type map was developed for the site (as shown in Figure 19.2, Wu et al., 1996), and much of these data were processed for analysis using Geographic Information System (GIS) technologies (Senay et al., 1998).

Low Altitude, Small Film Format Data

The Van Wert County area was flown approximately six times before and during the growing season of 1997. A single engine, six place aircraft was supplied by the local airport flight service for these flights of approximately 45 minutes duration. One or two photographers collected photographs from approximately 1,000 feet of altitude above ground level (AGL). Each used a single lens reflex type 35 mm camera of general manufacture. The optimal lens was a zoom telephoto with a focal length of approximately 100–180 mm or similar sort of lens. The film types included: black and white film for interpretation and publication, color slide film for interpretation and presentation to audiences, color infrared film for interpretation, and color print film for interpretation and for quick processing and printing to support immediate analyses (Lyon, 1993). The color infrared film was taken at ASA 100 to 125, using a yellow (minus blue light) filter and processed by Rocky Mountain Films in Aurora, CO. The color print and slide films were approximately 100 ASA and were taken with a polarizing filter to reduce atmospheric haze.

In 1998, a study which focused on evaluating crop responses to spatial variability was conducted at the Ohio MSEA site. As part of this study digital infrared data was obtained in May, August, and September. This information has a resolution of about 0.3 m by 0.3 m and was obtained from commercial services which use cameras mounted on aircraft flying at low altitude. We also used an AgLeader yield monitor to obtain on-the-go yield information. Scouting was performed every 1–4 weeks and extensive crop and soil data were obtained on several occasions at 27 to 36 sites for each of the three farming systems which were located on 9 ha fields.

Field Data Collection

The methods to collect and analyze field data are described in a number of publications cited here. In general, the following variables can be measured in the field and related to remotely sensed measures of light. They include: wet and/or dry biomass, crop residue by field scouting method or by removal and weighing, leaf area index (LAI), number of plant stocks in rows, areal cover of crop canopy, concentration or quantity of chlorophyll, crop height, plant moisture content, soil moisture content, soil color or reflectance, yield after harvesting, on-the-go yield, soil organic matter content, soil texture, soil fertility, soil nitrogen, soil phosphorus, soil potassium, presence and location of weeds, presence or location of wet areas during the growing season, presence of sandy or rocky or salty or other low quality soils, and related measures.

RESULTS

Crop Characteristics

All three experiments have demonstrated the capabilities of remote sensor measurements to supply information about crops. Whether the sensor was deployed in space, or in an expensive aircraft or in a relatively low-cost aircraft the analyses yielded information.

Remote sensing methods were useful in measuring the presence of wet biomass and dry biomass. The results include very good relationships between wet biomass and satellite and aircraft measured light (Price, 1992, 1993; Thenkabail et al., 1992, 1994; Van Deventer et al., 1997; Seney et al., 2000a, 2000b), and with dry biomass (Thenkabail et al., 1992).

A particularly valuable analysis approach demonstrated by our work and the work of a number of researchers was the use of ratio indices of remote sensed variables as compared to ground measured variables (Eidenshenk and Haas, 1992; Qi et al., 1993; Hatfield and Pinter, 1993; Lyon et al., 1998; Lunetta and Elvidge, 1998; Lyon, 2000, 2001). The most famous and widely used examples are those that employ the division of the red portion of the spectrum with that of the near infrared (Thenkabail et al., 1992; Price, 1994; Lyon et al., 1998). The normalized difference vegetation index (NDVI) has yielded very good results in our work with wet biomass (Thenkabail et al., 1992, 1994) and other ground measures (Van Deventer et al., 1997).

Other ratio indices have been tested by us (Thenkabail et al., 1992, 1994; Van Deventer et al., 1997; Lyon et al., 1998; Senay et al., 2000a, 2000b) and by others. Using combinations of middle infrared (TM5), red (TM3), and near infrared (TM4) provided information on crop moisture conditions or "stress" [e.g., STVI, Landsat (TM5 x TM3)/TM4,Thenkabail et al. 1992]. However, the NDVI is the most commonly used index and has been found to be the optimal example in several studies (Thenkabail et al., 1992; Lyon et al., 1998).

Yield and Crop Stresses

The low-altitude and low relative cost method of flying 35 mm photographs has demonstrated very good detection of spring, perennial weed growths. The optimal method is to use color print film and a 35 mm camera with a local aircraft service (Figure 19.3), and results are available within 24 hours or less of the flight. This method allows for a timely flight when weed conditions are obvious as compared to newly emerged or newly germinated crop, and provides adequate time to develop a plan and execute spraying of weeds. A similar approach can be used with infrared film but processing usually takes several days or longer (Figure 19.4a). Alternatively, digital cameras can be used and the data downloaded directly to a personal computer for near real-time analyses.

Remote sensing has the potential to be used to determine yield variability. Our earlier studies with Landsat TM data often showed low correlation between multispectral data and yields (Thenkabail et al., 1992, 1994). However, in the recent experiment at the MSEA site there was a strong linear correlation between yields and the near-infrared bands (Senay et al., 1998). It is possible that this improved ability to use spectral data to determine crop yields is associated with the higher resolution (1 m compared to 30 m) and with the availability of on-the-go yield data for use in developing relationships between yields and the spectral data.

Whether these types of relationships can be used to accurately determine yields in fields where on-the-go yield data are not available remains to be seen. However, it appears that even low cost infrared data obtained from a camera mounted in an aircraft will provide useful information on relative yield and biomass differences within a field. Figure 19.4b shows reduced crops in areas with high weed pressures (Canada Thistle and Giant Ragweed) that are presented in Figure 19.4a.

Figures 19.5a,b,c and d, show yield data and infrared images of a cornfield at the MSEA site during different growth stages. A well is located in the largest bare area, which is connected to an access road, and a power pylon is located in the other bare area (lower corner). This set of figures illustrates both the usefulness and the difficulties in using remotely sensed data. The August image suggests that there is little spatial variability in the crop characteristics. This image was taken when mature plants had a high moisture content and were fairly green. In contrast, the September image depicts senescent plants and considerable spatial variability is evident. Unlike the information presented in Figures 19.4a and 19.4b the variable crop responses in this field are primarily due to soil differences. In particular the soils in the southern (lower) part of the field are coarser, have lower soil water retention characteristics, and are less fertile than soils in other parts of the field. In Figure 19.5b, the small green patches in the middle of the field and near the bottom right-hand corner are areas with low plant densities. They correspond well to the lowest yielding (dark green) patches in Figure 19.5d.

Residue

We have used satellite Landsat TM data and high resolution multispectral data to study residue. On a field basis, in the northern Ohio study, we were able to differentiate between contrasting soil plains, and conventional versus conservation tillage (Van Deventer et al., 1997). However, we re-

a.

b.

Figure 19.3. Low altitude aerial photographs of farm fields in Van Wert County showing weeds in a field, and poor management resulting in low seeding, germination, or fertilizer applications in some rows.

cently tested the same linear logistic regression model on the Maquoketa River Watershed in Iowa and there was poor agreement between remote sensed based estimates and those from field surveys (P. Gowda, personal communication).

Soil Characteristics

A valuable result of the use of remote sensor data is in the evaluation of soil moisture conditions. In particular, the middle infrared (approximately 1.5 and 2.2 μm) and to a certain extent the near infrared (approximately 0.7 to 1.1 μm) supply good detail as to relative soil moisture conditions (Huete and Tucker, 1991; Van Deventer et al., 1997; Seney et al., 2000a, 2000b).

Soil fertility is an important issue in precision agriculture. Variables related to soil fertility are often not measured directly with a remote sensor, but are estimated from remote sensor variables using a related or surrogate variable (Lyon and McCarthy, 1995). Relative soil moisture and soil organic matter of bare soils can be measured directly. Relative soil phosphorus is measured in the field and compared to a surrogate variable measured remotely such as crop "greeness" or biomass, or leaf area index (Blackmer and Schepers, 1996).

Greater spectral resolution capabilities have allowed more detailed work on soil characteristics using indices similar to those previously published. Slightly modified indices that use the increased capabilities of Multispectral Scanner (MSS) aircraft data have proved useful. For example, the Simple Normalized Difference (SND) of the middle infrared (1.55–1.75 μm) divided by the near infrared (NIR, 0.92–1.10) which is similar to the Simple Vegetation Index (SVI) (Senay et al., 1998), and the Normalized Difference (ND) index using the equation (MIR-NIR)/(NIR + MIR) which is similar to NDVI in formulation (Seney et al. 2000a), have demonstrated increased capabilities. Both the SND and ND showed strong correlations with combined water in the soil (0.5 to 2.0 cm in depth) and in plants for corn (Senay et al., 2000a, 2000b).

Water in a Crop Environment

Most of our research has included an emphasis on water resource characteristics as they influence crops, soils, and water quality. These efforts have involved analyses of surface soil moisture, near surface drainage conditions, and vadose zone and groundwater conditions in support of watershed modeling (Ward and Elliot, 1995; Gowda et al., 1999). Remotely sensed measures have been valuable for a number of analyses. These include use of remote sensors or photography to document: wet soil conditions that may influence planting or germination; remote measurements of the impact of drought (Thenkabail et al., 1992); identification of land cover types in watersheds to help parameterize watershed models including GIS-based models (Moore et al., 1993; Gowda et al., 1999); calculation of watershed elevations for modeling (Garbrecht and Martz, 1993); remote measures of soil surface moisture (Van Deventer et al., 1997; Senay et al., 2000a, 2000b); use of remote sensing and GIS for nonpoint source modeling (Jakubauskas et al., 1992; Kang and Bartholic, 1994; Gowda et al., 1999); the accuracy of the remote sensor data or GIS products (Bolstad and Smith, 1995; Congalton and Green, 1998); and others.

Low altitude, 35 mm camera data can also identify drainage problems and the presence of subsurface drainage (Figure 19.6). During an early season overflight in Van Wert County, the weather and hydrology conditions were favorable for identification. Drainage systems were evident as light-toned areas which experienced better drainage than adjacent wet soil areas. The overflight occurred after a particularly wet spring, and about two days after the last rain. The drained areas were light toned because of the low relative moisture and hence increased reflectance as compared to adjacent, wetter soil areas (Figure 19.6).

General Characteristics of Fields

Statistical procedures and GIS technologies can be used to categorize spectral data into classes showing different crop characteristics, soils, topography and/or hydrological conditions. Figure 19.1 (see color section) shows more than 30 classes for the MSEA site which are based on the fine resolution visible, near infrared and middle infrared and elevation difference from the Digital Elevation Model (DEM). The classes themselves include measures of different levels (stalking, bio-

Figure 19.6. Low altitude aerial photographs of farm fields in Van Wert county showing wet areas, and the influence of subsurface drainage as lighter-toned linear features.

mass) of crop, differences in soils and elevation, and difference in plant and soil moisture conditions. The soybean crop is senescent and the corn is mature. The variety of spectrally distinct classes shows that a lot of detail can be sensed with the MSS sensor at one meter resolution and the additional discrimination power supplied by the data on relative elevation. Figure 19.7b is a false color composite of the fine resolution data and shows many of the same features that can be seen in Figure 19.1 (see color section). However, it provides less information on why these features occurred. These sorts of products can make a great amount of detail more obvious, and that same quantity of detail can be analyzed and evaluated as to accuracy (Lyon and McCarthy, 1995; Congalton and Green, 1998).

DISCUSSION AND CONCLUSIONS

The value of remotely sensed measurements lies in the utility of the data to identify normal and abnormal conditions in fields, and to do so rapidly over large areas. Producers know the characteristics of their fields, but things sometimes change rapidly and additional within-field data of a quantitative nature can only help the thoughtful management of crops and fields. The advent of Differential GPS positioning (Blackmer and Schepers, 1996) and on-the-go yield monitoring and more soil sampling allows the collection of powerful data sets. Combining these measurements with remotely sensed data and processing it all with GIS maps and modeling bodes well for detailed field management at low relative costs. Remotely sensed data can take the form of very low cost data either from shared satellite data collection and analysis efforts, or from low cost and rapid aerial photography from available local aircraft or commercial services (Lyon et al., 1986, 1995).

The best way to get started using precision agriculture technologies is to use some of the technologies to solve a specific field problem. The problem may be weeds, or variable fertility of soils, or poor germination. Use the technologies available to identify the locations of problems within the field and devise a solution and apply the solution. It is best to conduct the effort over several years to be sure that the result is not a chance situation with weather or some practice conducted in one certain year.

Infrared, color, or multispectral data acquired a few weeks after germination should provide information on weeds, soil spatial variability, and crop responses to this variability. Remote sensing data acquired during senescence is useful in identifying variability in crop densities and biomass. Data acquired during these two times and at other times during the growing season is also helpful in determining what caused the variability in crop responses.

Many experiments to characterize the potential of a given technology tend to be relatively expensive (Falkner, 1995; Lyon et al., 1995). As the technology and its utility become known, the researcher and industry then seeks to deliver this knowledge or technology at a cost of a few dollars per hectare. If the solution improves the yield in absolute terms, or removes the year to year variability of field yields, costs of this magnitude are economically viable.

A variety of results reported here or elsewhere have demonstrated that available remote sensor and GIS technologies can supply good information for management of crops, soils and waters on a within-field basis. Precision agriculture approaches can be implemented using a suite of these technologies, over time, to potentially improve yields and maintain soil and water quality.

ACKNOWLEDGMENTS

The authors wish to thank a number of people and sponsors for assisting in the development of these results.

The U.S. Environmental Protection Agency has helped to supply funding and aircraft remote sensor data through the National Exposure Research Laboratory Division at Research Triangle Park, NC and Las Vegas, NV, and through the assistance of EPA scientists Ross Lunetta and Clay Lake.

Some of the analysis of aerial photographs and farm conditions was conducted as a student Service Learning project at OSU, as part of a larger national effort called Learn and Serve America by the Ohio Campus Compact via the national Campus Compact at Brown University, as funded by the Corporation for National and Community Service.

Our thanks also go to our Departments at Ohio State University, OSU Extension, Van Wert County air service, Larry Lotz, Nathan Watermeier, Prasanna Gowda, and Jay Johnson.

BIBLIOGRAPHY

Blackmer, T., and J. Schepers, 1996. Using GPS to improve corn production and water quality. *GPS World*, March, pp. 44–52.

Bolstad, P., and J. Smith, 1995. Errors in GIS, assessing spatial data accuracy. In Lyon, J. and J. McCarthy, 1995.

Congalton, R., and K. Green, 1998. *Accuracy Assessment of Remotely Sensor Data: Principles and Practices*. CRC/Lewis Publishers, Boca Raton, FL.

Eidenshink, J., and R. Haas, 1992. Analyzing vegetation dynamics of land system with satellite data. *Geocarto International*, 1:53–61.

Falkner, E., 1995. *Aerial Mapping Methods and Applications*. Lewis/CRC Publishers, Boca Raton, FL, 352 pp.

Garbrecht, J., and L. Martz, 1993. Network and subwatershed parameters extracted from digital elevation models: The Bills creek experience. *Water Resources Bulletin,* 29:909–916.

Gowda, P., A. Ward, D. White, J. Lyon, and E. Desmond, 1999. The sensitivity of ADAPT model predictions of streamflows to parameters used to define hydrologic response units. *Transactions of the American Society for Agricultural Engineers,* 42:381–389.

Hatfield, J., and P. Pinter, 1993. Remote sensing for crop protection. *Crop Protection,* 12:403–413.

Huete, A., and C. Tucker, 1991. Investigation of soil influences in AVHRR red and near-infrared vegetation index imagery. *International Journal of Remote Sensing,* 12:1223–1242.

Jakubauskas, M., J. Whistler, M. Dillworth, and E. Martinko, 1992. Classifying remotely sensed data for use in an agricultural nonpoint source pollution model. *Journal of Soil and Water Conservation,* 47:179–183.

Kang, Y., and J. Bartholic, 1994. A GIS-based agricultural nonpoint source pollution management system at the watershed level. *Proceedings of the Annual ACSM/ASPRS Convention,* Reno, NV, pp. 281–289.

Lunetta, R., and C. Elvidge, 1998. *Remote Sensing Change Detection.* Ann Arbor Press, Chelsea, MI.

Lyon, J., 1993. *Practical Handbook for Wetlands Identification and Delineation.* CRC Lewis Publishers, Boca Raton, FL, 157 pp.

Lyon, J., 2003. *GIS for Water Resources and Watershed Management.* Taylor & Francis, London.

Lyon, J., and J. McCarthy, 1995. *Wetland and Environmental Applications of GIS.* Lewis Publishers, Boca Raton, FL.

Lyon, J., E. Falkner, and W. Bergen, 1995. Cost estimating photogrammetric and aerial photography services. *Journal of Surveying Engineering,* 121:63–86.

Lyon, J., J. McCarthy, and J. Heinen, 1986. Video digitization of aerial photographs for measurement of wind erosion damage on converted rangeland. *Photogrammetric Engineering and Remote Sensing,* 52:373–377.

Lyon, J., D. Yuan, R. Lunetta, and C. Elvidge, 1998. A change detection experiment using vegetation indices. *Photogrammetric Engineering and Remote Sensing,* 64:143–150.

Messer, J., R. Linthurst, and W. Overton, 1991. An EPA program for monitoring ecological status and trends. *Environmental Monitoring and Assessment,* 17:67–78.

Moore, I., A. Turner, J. Wilson, S. Jenson, and L. Band, 1993. GIS and land surface-subsurface process modeling. In M. Goodchild, et al., Eds., *GIS for Modeling,* Oxford University Press, Oxford, UK, pp.196–230.

Moran, M., Y. Inoue, and E. Barnes, 1997. Opportunities and limitations for image-based remote sensing in precision crop management. *Remote Sensing of Environment,* 61:319–346.

Price, J., 1992. Estimating vegetation amount from visible and near infrared reflectances. *Remote Sensing of Environment,* 41:29–34.

Price, J., 1993. Estimating leaf area index from satellite data. *IEEE Transactions on Geoscience and Remote Sensing,* 31:727–734.

Qi, J., A. Huete, M. Moran, A. Chehbouni, and R. Jackson, 1993. Interpretation of vegetation indices derived from multi-temporal SPOT images. *Remote Sensing of Environment,* 44:89–101.

Senay, G., A. Ward, J. Lyon, N. Fausey, and S. Nokes, 1998. Manipulation of high spatial resolution aircraft remote sensing data for use in site-specific farming. *Transactions of the American Society of Agricultural Engineers,* 41:489–495.

Senay, G., J. Lyon, A. Ward, and S. Nokes, 2000a. Using high spatial resolution multispectral data to classify corn and soybean crops. *Photogrammetric Engineering and Remote Sensing,* 66:319–327.

Senay, G., A. Ward, J. Lyon, N. Fausey, and L. Brown, 2000b. The relations between spectral data and water in a crop production environment. *International Journal of Remote Sensing,* 21:1897–1910.

Singh, A., 1989. Digital change detection techniques using remotely-sensed data. *International Journal of Remote Sensing,* 10:989–1003.

Thenkabail, P., A. Ward, J. Lyon, and C. Merry, 1994. Thematic Mapper vegetation indices for determining soybean and corn growth parameters. *Photogrammetric Engineering and Remote Sensing,* 60:437–442.

Thenkabail, P., A. Ward, J. Lyon, and P. Van Deventer, 1992. Landsat Thematic Mapper data (TM) indices for evaluating management and growth characteristics of soybeans and corn. *Transactions of the American Society of Agricultural Engineers,* 35:1441–1448.

Van Deventer, P., A. Ward, P. Gowda, and J. Lyon, 1997. Using Thematic Mapper data to identify contrasting soil plains and tillage practices. *Photogrammetric Engineering and Remote Sensing,* 63:87–93.

Van Sickle, J., 2001. *GPS for Land Surveyors.* Ann Arbor Press, Chelsea, MI.

Ward, A., and W. Elliot, 1995. *Environmental Hydrology.* Lewis Publishers, Boca Raton, FL.

Wu, Q., A. Ward, and S. Workman, 1996. Using GIS in simulation of nitrate leaching from heterogeneous unsaturated soils. *Journal of Environmental Quality,* 25:526–534.

Yang, C., G. Anderson, and J. Everitt, 1998. A view from above: Characterizing plant growth with aerial videography. *GPS World,* April, pp. 34–37.

Index